T0234113

# Modeling Tools
for
# Environmental
# Engineers
and
# Scientists

# Modeling Tools
## for
# Environmental
# Engineers
## and
# Scientists

## N. Nirmalakhandan

**CRC Press**
Taylor & Francis Group
Boca Raton  London  New York

CRC Press is an imprint of the
Taylor & Francis Group, an **informa** business

CRC Press LLC
Taylor & Francis Group
6000 Broken Sound Parkway NW, Suite 300
Boca Raton, FL 33487-2742

First issued in paperback 2020

© 2002 by Taylor & Francis Group, LLC
CRC Press is an imprint of Taylor & Francis Group, an Informa business

No claim to original U.S. Government works

ISBN 13: 978-0-367-57870-1 (pbk)
ISBN 13: 978-1-56676-995-2 (hbk)

**Visit the Taylor & Francis Web site at**
**http://www.taylorandfrancis.com**

**and the CRC Press Web site at**
**http://www.crcpress.com**

## Library of Congress Cataloging-in-Publication Data

Nirmalakhandan, N.
    Modeling tools for environmental engineers and scientists / N. Nirmalakhandan.
        p. cm.
    Includes bibliographical references and index.
    ISBN 1-56676-995-7
    1. Environmental sciences—Mathematical models. 2. Environmental
engineering—Mathematical models. I. Title.

    GE45.M37 K43 2001
    628′.01′1—dc21                                          2001052467

Library of Congress Card Number 2001052467

# Contents

## APPLICATIONS

# Preface

THIS book is not a treatise on environmental modeling. Several excellent books are currently available that do more than justice to the science of environmental modeling. The goal of this book is to bridge the gap between the science of environmental modeling and working models of environmental systems. More specifically, the intent of this book is to bring computer-based modeling within easy reach of subject matter experts and professionals who have shied away from modeling, daunted by the intricacies of computer programming and programming languages.

In the past two decades, interest in computer modeling in general and in environmental modeling in particular, have grown significantly. The number of papers and reports published on modeling, the number of specialty conferences on modeling held all over the world, and the number of journals dedicated to modeling efforts are evidence of this growth. Several factors such as better understanding of the underlying science, availability of high performance computer facilities, and increased regulatory concerns and pressures have fueled this growth. Scrutiny of those involved in environmental modeling, however, reveals that only a small percentage of experts are active in the modeling efforts; namely those who also happen to be skilled in computer programming.

For the rest of us, computer modeling has remained a challenging task until recently. A new breed of software authoring packages has now become available that enables nonprogrammers to develop their own models without having to learn programming languages. These packages feature English-like syntax and easy-to-use yet extremely powerful mathematical, analytical, computational, and graphical functions, and user-friendly interfaces. They can drastically reduce the time, effort, and programming skills required to develop professional quality user-friendly models. This book describes eight such software packages, and, with over 50 modeling examples, illustrates how they can be adapted for almost any type of modeling project.

The contents of this book are organized into nine chapters. Chapter 1 is an introduction to modeling. Chapter 2 focuses on the science and art of mathematical modeling. Chapter 3 contains a primer on mathematics with examples of computer implementation of standard mathematical calculi. Chapters 4, 5, and 6 contain reviews of the fundamentals of environmental processes, engineered systems, and natural systems, respectively. Chapter 7 describes and compares the eight software packages selected here for developing environmental models. Chapters 8 and 9 are devoted entirely to modeling examples covering engineered and natural systems, respectively.

The book can be of benefit to those who have been yearning to venture into modeling, as well as to those who have been using traditional language-based approaches for modeling. In the academic world, this book can be used as the main text to cultivate computer modeling skills at the freshman or junior levels. It can be used as a companion text in "fate and transport" or "environmental modeling" types of courses. It can be useful to graduate students planning to incorporate some form of modeling into their research. Faculty can benefit from this book in developing special purpose models for teaching, research, publication, or consulting. Practicing professionals may find it useful to develop custom models for limited use, preliminary analysis, and feasibility studies. Suggested uses of this book by different audiences at different levels are included in Chapter 1.

As a final note, this book should not be taken as a substitute for the user manuals that accompany software packages; rather, it takes off from where the user manuals stop, and demonstrates the types of finished products (models, in this case) that can be developed by integrating the features of the software. This book can, perhaps, be compared to a road map, which can take nonprogrammers from the problem statement to a working computer-based mathematical model. Different types of vehicles can be used for the journey. The intention here is not to teach how to drive the vehicles but rather to point out the effort required by the different vehicles, their capabilities, advantages, disadvantages, special features, and limitations.

# Acknowledgments

I would like to acknowledge several individuals who, directly or indirectly, were responsible for giving me the strength, confidence, opportunity, and support to venture into this project. First, I would like to recognize all my teachers in Sri Lanka, some of whom probably did not even know me by name. Yet, they gave me a rigorous education, particularly in mathematics, for which I am forever thankful. I also want to acknowledge the continuing encouragement, guidance, motivation, and friendship of Professor Richard Speece, to whom I gratefully attribute my academic career. The five years that I shared with Professor Speece have been immensely satisfying and most productive. I want to thank my department head, Professor Kenneth White, for supporting my efforts throughout my academic tenure, without which I could not have found the time or the resources to embark on this project. Thanks to my departmental colleagues and the administration at New Mexico State University for allowing me to go on sabbatical leave to complete this project.

The donation of full versions of software packages by MathSoft, Inc., The Mathworks, Inc., Universal Technical Systems, Inc., Wolfram Research, Inc., and High Performance Systems, Inc., is gratefully acknowledged. The support, encouragement, and patience of Dr. Joseph Eckenrode and his editorial staff at Technomic Publishing Co., Inc. (now owned by CRC Press LLC) is fully appreciated.

I am indebted to my grandparents and to my parents for the sacrifices they made to ensure that I received the finest education from the earliest age; and to the families of my sister and sister-in-law for their support and compassion, particularly during my graduate studies. Finally, I am thankful to my wife Gnana and our sons Rajeev and Sanjeev, for their support, understanding, and tolerance throughout this and many other projects, which took much of my time and attention away from them. Sanjeev also contributed to this work directly in setting up some of the models and by sharing some of the agonies and ecstasies of computer programming.

# PART I

# Fundamentals

## CHAPTER 1

# Introduction to Modeling

## CHAPTER PREVIEW

*In this chapter, an overview of the process of modeling is presented. Different approaches to modeling are identified first, and features of mathematical modeling are detailed. Alternate classifications of mathematical models are addressed. A case history is presented to illustrate the benefits and scope of environmental modeling. A road map through this book is presented, identifying the topics to be covered in the following chapters and potential uses of the book.*

## 1.1 WHAT IS MODELING?

**M**ODELING can be defined as the process of application of fundamental knowledge or experience to simulate or describe the performance of a real system to achieve certain goals. Models can be cost-effective and efficient tools whenever it is more feasible to work with a substitute than with the real, often complex systems. Modeling has long been an integral component in organizing, synthesizing, and rationalizing observations of and measurements from real systems and in understanding their causes and effects.

In a broad sense, the goals and objectives of modeling can be twofold: research-oriented or management-oriented. Specific goals of modeling efforts can be one or more of the following: to interpret the system; to analyze its behavior; to manage, operate, or control it to achieve desired outcomes; to design methods to improve or modify it; to test hypotheses about the system; or to forecast its response under varying conditions. Practitioners, educators, researchers, and regulators from all professions ranging from business to management to engineering to science use models of some form or another in their respective professions. It is probably the most common denominator among all endeavors in such professions, especially in science and engineering.

The models resulting from the modeling efforts can be viewed as logical and rational representations of the system. A model, being a representation and a working hypothesis of a more complex system, contains adequate but less information than the system it represents; it should reflect the features and characteristics of the system that have significance and relevance to the goal. Some examples of system representations are verbal (e.g., language-based description of size, color, etc.), figurative (e.g., electrical circuit networks), schematic (e.g., process and plant layouts), pictographic (e.g., three-dimensional graphs), physical (e.g., scaled models), empirical (e.g., statistical models), or symbolic (e.g., mathematical models). For instance, in studying the ride characteristics of a car, the system can be represented verbally with words such as "soft" or "smooth," figuratively with spring systems, pictographically with graphs or videos, physically with a scaled material model, empirically with indicator measurements, or symbolically using kinematic principles.

Most common modeling approaches in the environmental area can be classified into three basic types—physical modeling, empirical modeling, and mathematical modeling. The third type forms the foundation for computer modeling, which is the focus of this book. While the three types of modeling are quite different from one another, they complement each other well. As will be seen, both physical and empirical modeling approaches provide valuable information to the mathematical modeling process. These three approaches are reviewed in the next section.

## 1.1.1 PHYSICAL MODELING

Physical modeling involves representing the real system by a geometrically and dynamically similar, scaled model and conducting experiments on it to make observations and measurements. The results from these experiments are then extrapolated to the real systems. Dimensional analysis and similitude theories are used in the process to ensure that model results can be extrapolated to the real system with confidence.

Historically, physical modeling had been the primary approach followed by scientists in developing the fundamental theories of natural sciences. These included laboratory experimentation, bench-scale studies, and pilot-scale tests. While this approach allowed studies to be conducted under controlled conditions, its application to complex systems has been limited. Some of these limitations include the need for dimensional scale-up of "small" systems (e.g., colloidal particles) or scale-down of "large" ones (e.g., acid rain), limited accessibility (e.g., data collection); inability to accelerate or slow down processes and reactions (e.g., growth rates), safety (e.g., nuclear reactions), economics (e.g., Great Lakes reclamation), and flexibility (e.g., change of diameter of a pilot column).

## 1.1.2 EMPIRICAL MODELING

Empirical modeling (or black box modeling) is based on an inductive or data-based approach, in which past observed data are used to develop relationships between variables believed to be significant in the system being studied. Statistical tools are often used in this process to ensure validity of the predictions for the real system. The resulting model is considered a "black box," reflecting only *what* changes could be expected in the system performance due to changes in inputs. Even though the utility value of this approach is limited to predictions, it has proven useful in the case of complex systems where the underlying science is not well understood.

## 1.1.3 MATHEMATICAL MODELING

Mathematical modeling (or mechanistic modeling) is based on the deductive or theoretical approach. Here, fundamental theories and principles governing the system along with simplifying assumptions are used to derive mathematical relationships between the variables known to be significant. The resulting model can be calibrated using historical data from the real system and can be validated using additional data. Predictions can then be made with predefined confidence. In contrast to the empirical models, mathematical models reflect *how* changes in system performance are related to changes in inputs.

The emergence of mathematical techniques to model real systems have alleviated many of the limitations of physical and empirical modeling. Mathematical modeling, in essence, involves the transformation of the system under study from its natural environment to a mathematical environment in terms of abstract symbols and equations. The symbols have well-defined meanings and can be manipulated following a rigid set of rules or "mathematical calculi." Theoretical concepts and process fundamentals are used to derive the equations that establish relationships between the system variables. By feeding known system variables as inputs, these equations or "models" can then be solved to determine a desired, unknown result. In the precomputer era, mathematical modeling could be applied to model only those problems with closed-form solutions; application to complex and dynamic systems was not feasible due to lack of computational tools.

With the growth of high-speed computer hardware and programming languages in the past three decades, mathematical techniques have been applied successfully to model complex and dynamic systems in a computer environment. Computers can handle large volumes of data and manipulate them at a minute fraction of the time required by manual means and present the results in a variety of different forms responsive to the human mind. Development of computer-based mathematical models, however, remained a demanding task within the grasp of only a few with subject-matter expertise and computer programming skills.

During the last decade, a new breed of software packages has become available that enables subject matter experts with minimal programming skills to build their own computer-based mathematical models. These software packages can be thought of as tool kits for developing applications and are sometimes called software authoring tools. Their functionality is somewhat similar to the following: a web page can be created using hypertext marking language (HTML) directly. Alternatively, one can use traditional word-processing programs (such as Word®[1]), or special-purpose authoring programs (such as PageMill®[2]), and click a button to create the web page without requiring any knowledge of HTML code.

Currently, several different types of such syntax-free software authoring tools are commercially available for mathematical model building. They are rich with built-in features such as a library of preprogrammed mathematical functions and procedures, user-friendly interfaces for data entry and running, post-processing of results such as plotting and animation, and high degrees of interactivity. These authoring tools bring computer-based mathematical modeling within easy reach of more subject matter experts and practicing professionals, many of whom in the past shied away from it due to lack of computer programming and/or mathematical skills.

## 1.2 MATHEMATICAL MODELING

The elegance of mathematical modeling needs to be appreciated: a single mathematical formulation can be adapted for a wide number of real systems, with the symbols taking on different meanings depending on the system. As an elementary example, consider the following linear equation:

$$Y = mX + C \tag{1.1}$$

The "mathematics" of this equation is very well understood as is its "solution." The readers are probably aware of several real systems where Equation (1.1) can serve as a model (e.g., velocity of a particle falling under gravitational acceleration or logarithmic growth of a microbial population). As another example, the partial differential equation

$$\frac{\partial \phi}{\partial t} = \alpha \frac{\partial^2 \phi}{\partial x^2} \tag{1.2}$$

---

[1]Word® is a registered trademark of Microsoft Corporation. All rights reserved.
[2]PageMill® is a registered trademark of Adobe Systems Incorporated. All rights reserved.

can model the temperature profile in a one-dimensional heat transfer problem or the concentration of a pollutant in a one-dimensional diffusion problem. Thus, subject matter experts can reduce their models to standard mathematical forms and adapt the standard mathematical calculi for their solution, analysis, and evaluation.

Mathematical models can be classified into various types depending on the nature of the variables, the mathematical approaches used, and the behavior of the system. The following section identifies some of the more common and important types in environmental modeling.

### 1.2.1 DETERMINISTIC VS. PROBABILISTIC

When the variables (in a static system) or their changes (in a dynamic system) are well defined with certainty, the relationships between the variables are fixed, and the outcomes are unique, then the model of that system is said to be deterministic. If some unpredictable randomness or probabilities are associated with at least one of the variables or the outcomes, the model is considered probabilistic. Deterministic models are built of algebraic and differential equations, while probabilistic models include statistical features.

For example, consider the discharge of a pollutant into a lake. If all of the variables in this system, such as the inflow rate, the volume of the lake, etc., are assumed to be average fixed values, then the model can be classified as deterministic. On the other hand, if the flow is taken as a mean value with some probability of variation around the mean, due to runoff, for example, a probabilistic modeling approach has to be adapted to evaluate the impact of this variable.

### 1.2.2 CONTINUOUS VS. DISCRETE

When the variables in a system are continuous functions of time, then the model for the system is classified as continuous. If the changes in the variables occur randomly or periodically, then the corresponding model is termed discrete. In continuous systems, changes occur continuously as time advances evenly. In discrete models, changes occur only when the discrete events occur, irrespective of the passage of time (time between those events is seldom uniform). Continuous models are often built of differential equations; discrete models, of difference equations.

Referring to the above example of a lake, the volume or the concentration in the lake might change with time, but as long as the inflow remains non-zero, the system will be amenable to continuous modeling. If random events such as rainfall are to be included, a discrete modeling approach may have to be followed.

### 1.2.3 STATIC VS. DYNAMIC

When a system is at steady state, its inputs and outputs do not vary with passage of time and are average values. The model describing the system under those conditions is known as static or steady state. The results of a static model are obtained by a single computation of all of the equations. When the system behavior is time-dependent, its model is called dynamic. The output of a dynamic model at any time will be dependent on the output at a previous time step and the inputs during the current time step. The results of a dynamic model are obtained by repetitive computation of all equations as time changes. Static models, in general, are built of algebraic equations resulting in a numerical form of output, while dynamic models are built of differential equations that yield solutions in the form of functions.

In the example of the lake, if the inflow and outflow remain constant, the resulting concentration of the pollutant in the lake will remain at a constant *value,* and the system can be modeled by a static model. But, if the inflow of the pollutant is changed from its steady state value to another, its concentration in the lake will change as a function of time and approach another steady state value. A dynamic model has to be developed if it is desired to trace the concentration profile during the change, as a *function* of time.

### 1.2.4 DISTRIBUTED VS. LUMPED

When the variations of the variables in a system are continuous functions of time and space, then the system has to be modeled by a distributed model. For instance, the variation of a property, $C$, in the three orthogonal directions $(x, y, z)$, can be described by a distributed function $C = f(x,y,z)$. If those variations are negligible in those directions within the system boundary, then $C$ is uniform in all directions and is independent of $x$, $y$, and $z$. Such a system is referred to as a lumped system. Lumped, static models are often built of algebraic equations; lumped, dynamic models are often built of ordinary differential equations; and distributed models are often built of partial differential equations.

In the case of the lake example, if mixing effects are (observed or thought to be) significant, then a distributed model could better describe the system. If, on the other hand, the lake can be considered completely mixed, a lumped model would be adequate to describe the system.

### 1.2.5 LINEAR VS. NONLINEAR

When an equation contains only one variable in each term and each variable appears only to the first power, that equation is termed *linear,* if not, it is known as *nonlinear.* If a model is built of linear equations, the model responses are additive in their effects, i.e., the output is directly proportional to the input, and outputs satisfy the principle of superpositioning. For instance, if an input $I_1$ to a system produces an output $O_1$, and another input $I_2$ produces an output of $O_2$, then a combined input of $(\alpha I_1 + \beta I_2)$ will produce an output of $(\alpha O_1 + \beta O_2)$. Superpositioning cannot be applied in nonlinear models.

In the lake example, if the reactions undergone by the pollutant in the lake are assumed to be of first order, for instance, then the linearity of the resulting model allows superpositioning to be applied. Suppose the input to the lake is changed from a steady state condition, then the response of the lake can be found by adding the response following the general solution (due to the initial conditions) to the response following the particular solution (due to the input change) of the differential equation governing the system.

## 1.2.6 ANALYTICAL VS. NUMERICAL

When all the equations in a model can be solved algebraically to yield a solution in a closed form, the model can be classified as analytical. If that is not possible, and a numerical procedure is required to solve one or more of the model equations, the model is classified as numerical.

In the above example of the lake, if the entire volume of the lake is assumed to be completely mixed, a simple analytical model may be developed to model its steady state condition. However, if such an assumption is unacceptable, and if the lake has to be compartmentalized into several layers and segments for detailed study, a numerical modeling approach has to be followed.

A comparison of the above classifications is summarized in Figure 1.1. Indicated at the bottom section of this figure are the common mathematical analytical methods appropriate for each type of model. These classifications are presented here to stress the necessity of understanding input data requirements, model formulation, and solution procedures, and to guide in the selection of the appropriate computer software tool in modeling the system. Most environmental systems can be approximated in a satisfactory manner by linear and time variant descriptions in a lumped or distributed manner, at least for specified and restricted conditions. Analytical solutions are possible for limited types of systems, while solutions may be elaborate or not currently available for many others. Computer-based mathematical modeling using numerical solutions can provide valuable insight in such cases.

The goal of this book is to illustrate, with examples, the application of a variety of software packages in developing computer-based mathematical models in the environmental field. The examples included in the book fall into the following categories: static, dynamic, continuous, deterministic (probabilistic, at times), analytical, numerical, and linear.

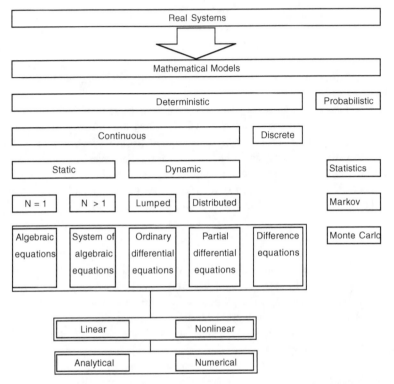

**Figure 1.1** Classification of mathematical models ($N$ = number of variables).

## 1.3 ENVIRONMENTAL MODELING

The application of mathematical modeling in various fields of study has been well illustrated by Cellier (1991). According to Cellier's account, such models range from the well-defined and rigorous "white-box" models to the ill-defined, empirical "black-box" models. With white-box models, it is suggested that one could proceed directly to design of full-scale systems with confidence, while with black-box models, that remains a speculative theory. A modified form of the illustration of Cellier is shown in Table 1.1.

Mathematical modeling in the environmental field can be traced back to the 1900s, the pioneering work of Streeter and Phelps on dissolved oxygen being the most cited. Today, driven mainly by regulatory forces, environmental studies have to be multidisciplinary, dealing with a wide range of pollutants undergoing complex biotic and abiotic processes in the soil, surface water, groundwater, ocean water, and atmospheric compartments of the ecosphere. In addition, environmental studies also encompass equally diverse engineered reactors and processes that interact with the natural environment

Table 1.1 Range of Mathematical Models

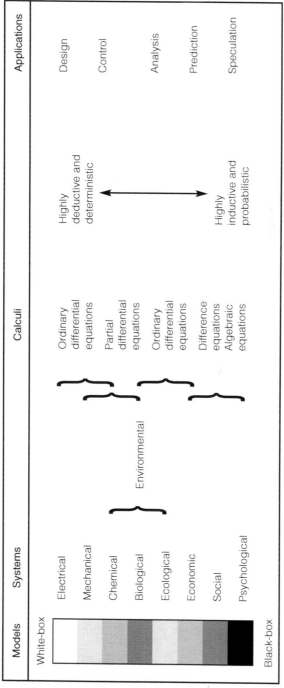

through several pathways. Consequently, modeling of large-scale environmental systems is often a complex and challenging task. The impetus for developing environmental models can be one or more of the following:

(1) To gain a better understanding of and glean insight into environmental processes and their influence on the fate and transport of pollutants in the environment

(2) To determine short- and long-term chemical concentrations in the various compartments of the ecosphere for use in regulatory enforcement and in the assessment of exposures, impacts, and risks of existing as well as proposed chemicals

(3) To predict future environmental concentrations of pollutants under various waste loadings and/or management alternatives

(4) To satisfy regulatory and statutory requirements relating to environmental emissions, discharges, transfers, and releases of controlled pollutants

(5) To use in hypothesis testing relating to processes, pollution control alternatives, etc.

(6) To implement in the design, operation, and optimization of reactors, processes, pollution control alternatives, etc.

(7) To simulate complex systems at real, compressed, or expanded time horizons that may be too dangerous, too expensive, or too elaborate to study under real conditions

(8) To generate data for post-processing, such as statistical analysis, visualization, and animation, for better understanding, communication, and dissemination of scientific information

(9) To use in environmental impact assessment of proposed new activities that are currently nonexistent

Above all, the formal exercise of designing and building a model may be more valuable than the actual model itself or its use in that the knowledge about the problem is organized and crystallized to extract the maximum benefit from the effort and the current knowledge about the subject. Typical issues and concerns in various environmental systems and the use of mathematical models in addressing them are listed in Appendix 1.1, showing the wide scope of environmental modeling.

The use of and need for mathematical models, their scope and utility value, and the computer-based approaches used in developing them can best be illustrated by a case history (Nirmalakhandan et al., 1990, 1991a, 1991b, 1992a, 1992b, 1993).

*Case History:* Improvement of the Air-Stripping Process

The air-stripping (A/S) process has been used in the chemical engineering field for over 50 years. During the early 1980s, this process was adapted by environmental engineers for remediating groundwaters contaminated with organic contaminants. While A/S is a cost-effective process for removing volatile organic contaminants (VOCs), its application at sites contaminated with semivolatile organic contaminants (SVOCs) had been limited by prohibitive energy requirements. This case study summarizes how mathematical modeling was utilized in developing and demonstrating a unique modification to the A/S process that had the potential for gaining significant improvements in applicability, efficiency, energy consumption, and overall treatment costs.

In the A/S process, the VOC-contaminated water is pumped to the top of a packed tower from where it flows down through the packing media under gravity. A countercurrent stream of air blown from the bottom of the tower strips the VOCs, and the treated water is collected at the bottom of the tower. From the mass transfer theory, it is known that the efficiency of the process and its applicability to SVOCs can be improved by increasing the driving force for mass transfer. The driving force can be improved by increasing the airflow rate. But, increasing the airflow rate will not only increase pressure drop and energy consumption, but it will also lead to process failure due to flooding of the tower.

It was hypothesized that if the airflow could be distributed along the depth of the packing, the driving force for mass transfer could be increased: the fresh air entering the tower along its depth will dilute the upcoming contaminated air through the tower, thus increasing the driving force throughout the full depth of the tower. At the same time, the overall pressure drop will not be that high. In combination, these two factors could be expected to reduce the packing depth requirements, pressure drop, and energy consumption, leading to reduced capital and operating costs.

To verify this hypothesis, a mathematical model based on fundamental mass theories was formulated. The model was then used to compare the conventional A/S process against the proposed process configuration under identical input conditions. This modeling exercise confirmed that the proposed configuration could result in a 50% reduction in packing depth, a 40% reduction in pressure drop, and a 40% reduction in energy requirement, for comparable removal efficiencies. Based on these model results, the American Water Works Association (AWWA) funded a research project to verify the hypothesis and validate the process model at pilot scale. The model was used to optimize the process and design the optimal pilot-scale process. This pilot-scale test confirmed the hypothesis and validated the model predictions. The validated model was then used to design a prototype-scale system and a field-scale system that produced results that were used to further validate the mathematical model over a wide range of operating variables under field conditions.

## 1.4 OBJECTIVES OF THIS BOOK

The premise of this book is that computer-based mathematical modeling is a powerful and essential tool appropriate for analyzing and evaluating a wide range of complex environmental applications. Desktop and laptop computers with computing capabilities approaching supercomputer-level performance are now available at reasonable prices. Several new powerful software packages have emerged that bring mathematical modeling within reach of nonprogrammers.

The primary objective of this book is to introduce several types of computer software packages that subject matter experts with minimal computer programming skills can use to develop their own models. Even though this book has the words "environmental modeling" in its title and is illustrated with examples in that area, professionals in other fields can also learn about the features of the various software packages identified, benefit from the examples included, and be able to adapt them in their own areas.

The motivation for a book of this nature is that the users manuals that accompany software often are not focused toward application of the software in modeling. In modeling, one has to integrate several features to achieve the modeling goal. In all fairness, the manuals are well written to illustrate the functioning of individual features of the software. They cannot be expected to include examples of integration of features for specific purposes. This book takes off from where the manuals end and provides examples of integrated applications of the software to specific problems.

Further impetus for this book stems from the fact that computer usage, particularly in modeling, analyzing, synthesizing, and simulating in engineering curricula, is being advocated by cognitive researchers as one of the effective ways of improving the teaching and learning processes. Recognizing this, the Accreditation Board of Engineering Technologies (ABET) is placing strong emphasis on computer usage in engineering curricula. Several engineering programs have recently initiated freshman undergraduate-level courses to introduce computer applications in engineering. This book includes basic mathematical concepts and scientific principles for use in such classes to cultivate modeling skills from early stages.

Chapter 2 of the book contains a discussion of the various steps involved in developing mathematical models. As pointed out in Chapter 2, modeling is part science and part art, and as such, modelers have developed their own style of accomplishing the task. While it is recognized that different approaches are being practiced, the steps and tasks identified in Chapter 2 are key elements in the process and have to be taken into consideration in some form or another. The procedures presented in Chapter 2 are in no way intended to be followed in every case.

Chapters 3, 4, 5, and 6 present reviews of fundamental concepts in mathematics, environmental processes, engineered environmental systems, and natural environmental systems, respectively. The material included in this book is not to be construed as a formal and/or complete treatment of the respective topics but as a *review* of the fundamentals involved, providing a basic starting point in the modeling process.

Chapter 7 contains a comparative presentation of various types of software authoring packages that can be used for developing mathematical models. These include spreadsheets, mathematical software, and dynamic modeling software. A total of six different packages are identified and compared using one common problem. Rather than attempt to include every software package commercially available, selected examples of software products representing the different types of systems are included.

Chapters 8 and 9 contain several examples of applications of the different types of software packages in modeling engineered and natural systems, respectively. Here, the main focus is on the use of the various packages in the computer implementation of the modeling process rather than on the fundamentals of the science behind the systems. However, a brief review of and references to related theories and principles are included, along with pertinent details on the model formulation process as a prerequisite to the computer implementation process.

Suggested uses of this book by various audiences are indicated in Appendix 1.2. It is hoped that the book will be able to serve as a primary textbook in environmental modeling courses; a companion book in unit operations or environmental fate and transport courses; a guidebook to faculty interested in modeling work for teaching, research, and publication; a tool for graduate students involved in modeling-oriented research; and a reference book for practicing professionals in the environmental area.

## APPENDIX 1.1 TYPICAL USES OF MATHEMATICAL MODELS

| Environmental Medium | Issues/Concerns | Use of Models in |
|---|---|---|
| Atmosphere | Hazardous air pollutants; air emissions; toxic releases; acid rain; smog; CFCs; particulates; health concerns; global warming | Concentration profiles; exposure; design and analysis of control processes and equipment; evaluation of management actions; environmental impact assessment of new projects; compliance with regulations |
| Surface water | Wastewater treatment plant discharges; industrial discharges; agricultural/urban runoff; storm water discharges; potable water source; food chain | Fate and transport of pollutants; concentration plumes; design and analysis of control processes and equipment; wasteload allocations; evaluation of management actions; environmental impact assessment of new projects; compliance with regulations |
| Groundwater | Leaking underground storage tanks; leachates from landfills and agriculture; injection; potable water source | Fate and transport of pollutants; design and analysis of remedial actions; drawdowns; compliance with regulations |
| Subsurface | Land application of solid and hazardous wastes; spills; leachates from landfills; contamination of potable acquifers | Fate and transport of pollutants; concentration plumes; design and analysis of control processes; evaluation of management actions |
| Ocean | Sludge disposal; spills; outfalls; food chain | Fate and transport of pollutants; concentration plumes; design and analysis of control processes; evaluation of management actions |

**APPENDIX 1.2 SUGGESTED USES OF THE BOOK**

| Chapter | Topics | Target Audience | | | | | | |
|---|---|---|---|---|---|---|---|---|
| | | Lower-level BS students | Upper-level BS and beginning MS students | Advanced-level MS and PhD students | Subject matter experts with minimal computer modeling expertise | Subject matter experts using language-based software | Subject matter experts who have used at least one authoring software program | Practicing engineers, project managers, regulators |
| 2 | Fundamentals of mathematical modeling | Read and review | Read and review | Read and review | Read and review | | | Review |
| 3 | Primer on mathematics | Read and review | Read and review | Review | Review | | | |
| 4 | Fundamentals of environmental processes | Read and review | Read and review | Review | | | | Review |
| 5 | Fundamentals of engineered systems | Read and review | Read and review | Review | | | | Review |
| 6 | Fundamentals of natural systems | Read and review | Read and review | Read and review | | | | Review |

**APPENDIX 1.2 (continued)**

| Chapter | Topics | Target Audience | | | | | | |
|---|---|---|---|---|---|---|---|---|
| | | Lower-level BS students | Upper-level BS and beginning MS students | Advanced-level MS and PhD students | Subject matter experts with minimal computer modeling expertise | Subject matter experts using language-based software | Subject matter experts who have used at least one authoring software program | Practicing engineers, project managers, regulators |
| 7 | Software for developing mathematical models | Read and review | Read and review | Read and review | Read and review | Read and review | Read and review | Read and review |
| 8 | Modeling of engineered systems | | Read and review | Read and review | Read and review | Read and review | Review | Review |
| 9 | Modeling of natural systems | | Read and review | Read and review | Read and review | Read and review | Review | Review |

## CHAPTER 2

# Fundamentals of Mathematical Modeling

## CHAPTER PREVIEW

*In this chapter, formal definitions and terminology relating to mathematical modeling are presented. The key steps involved in developing mathematical models are identified, and the tasks to be completed under each step are detailed. While the suggested procedure is not a standard one, it includes the crucial components to be addressed in the process. The application of these steps in developing a mathematical model for a typical environmental system is illustrated.*

## 2.1 DEFINITIONS AND TERMINOLOGY IN MATHEMATICAL MODELING

G ENERAL background information on models was presented in Chapter 1, where certain terms were introduced in a general manner. Before continuing on to the topic of developing mathematical models, it is necessary to formalize certain terminology, definitions, and conventions relating to the modeling process. Recognition of these formalities can greatly help in the selection of the modeling approach, data needs, theoretical constructs, mathematical tools, solution procedures, and, hence, the appropriate computer software package(s) to complete the modeling task. In the following sections, the language in mathematical modeling is clarified in the context of modeling of environmental systems.

### 2.1.1 SYSTEM/BOUNDARY

A "system" can be thought of as a collection of one or more related objects, where an "object" can be a physical entity with specific attributes or characteristics. The system is isolated from its surroundings by the "boundary," which can be physical or imaginary. (In many books on modeling, the term "environment" is used instead of "surroundings" to indicate everything outside the boundary; the reason for picking the latter is to avoid the confusion in the context of this book that focuses on modeling the environment. In other words, environment is the system we are interested in modeling, which is enclosed by the boundary.) The objects within a system may or may not interact with each other and may or may not interact with objects in the surroundings, outside the boundary. A system is characterized by the fact that the modeler can define its boundaries, its attributes, and its interactions with the surroundings to the extent that the resulting model can satisfy the modeler's goals.

The largest possible system of all, of course, is the universe. One can, depending on the modeling goals, isolate a part of the universe such as a continent, or a country, or a city, or the city's wastewater treatment plant, or the aeration tank of the city's wastewater treatment plant, or the microbial population in the aeration tank, and define that as a system for modeling purposes. Often, the larger the system, the more complex the model. However, the effort can be made more manageable by dissecting the system into smaller subsystems and including the interactions between them.

## 2.1.2 OPEN/CLOSED, FLOW/NONFLOW SYSTEMS

A system is called a closed system when it does not interact with the surroundings. If it interacts with the surroundings, it is called an open system. In closed systems, therefore, neither mass nor energy will cross the boundary; whereas in open systems, mass and energy can. When mass does not cross the boundary (but energy does), an open system may be categorized as a nonflow system. If mass crosses the boundary, it is called a flow system.

While certain batch processes may be approximated as closed systems, most environmental systems interact with the surroundings in one way or another, with mass flow across the boundary. Thus, most environmental systems have to be treated as open, flow systems.

## 2.1.3 VARIABLES/PARAMETERS/INPUTS/OUTPUTS

The attributes of the system and of the surroundings that have significant impact on the system are termed "variables." The term variable includes those attributes that change in value during the modeling time span and those that remain constant during that period. Variables of the latter type are often referred to as parameters. Some parameters may relate to the system, and some may relate to the surroundings.

A system may have numerous attributes or variables. However, as mentioned before, the modeler needs to select only those that are significant and relevant to the modeler's goal in the modeling process. For example, in the case of the aeration tank, its attributes can include biomass characteristics, volume of mixed liquor, its color, temperature, viscosity, specific weight, conductivity, reflectivity, etc., and the attributes of the surroundings may be flow rate, mass input, wind velocity, solar radiation, etc. Even though many of the attributes may be interacting, only a few (e.g., biomass characteristics, volume, flow rate, mass input) are identified as variables of significance and relevance based on the modeler's goals (e.g., the efficiency of the aeration tank).

Variables that change in value fall into two categories: those that are generated by the surroundings and influence the behavior of the system, and those that are generated by the system and impact the surroundings. The former are called "inputs," and the latter are called "outputs." In the case of the aeration tank, the mass inflow can be an input, the concentration leaving the tank, an output, and the volume of the tank, a parameter. In mathematical language, inputs are considered independent variables, and outputs are considered dependent variables. The inputs and model parameters are often known or defined in advance; they drive the model to produce some output. In the context of modeling, relationships are sought between inputs and outputs, with the parameters acting as model coefficients.

At this point, a very important factor has to be recognized; in the real system, not all significant and relevant variables and/or parameters may be accessible for control or manipulation; likewise, not all outputs may be accessible for observation or measurement. However, in mathematical models, all inputs and parameters are readily available for control or manipulation, and all outputs are accessible. It also follows that, in mathematical modeling, modelers can suppress "disturbances" that are unavoidable in the real systems. These traits are of significant value in mathematical modeling.

However, numerical values for the variables will be needed to execute the model. Some values are set by the modeler as inputs. Other system parameter data can be obtained from many sources, such as the scientific literature, experimentation on the real system or physical models, or by adapting estimation methods. Accounts of experimentation techniques and parameter estimation methods for determining such data can be found elsewhere and are beyond the scope of this book.

## 2.2 STEPS IN DEVELOPING MATHEMATICAL MODELS

The craft of mathematical model development is part science and part art. It is a multistep, iterative, trial-and-error process cycling through hypotheses formation, inferencing, testing, validating, and refining. It is common practice to start from a simple model and develop it in steps of increasing complexity, until it is capable of replicating the observed or anticipated behavior of the real system to the extent that the modeler expects. It has to be kept in mind that all models need not be perfect replicates of the real system. If all the details of the real system are included, the model can become unmanageable and be of very limited use. On the other hand, if significant and relevant details are omitted, the model will be incomplete and again be of limited use. While the scientific side of modeling involves the integration of knowledge to build and solve the model, the artistic side involves the making of a sensible compromise and creating balance between two conflicting features of the model: degree of detail, complexity, and realism on one hand, and the validity and utility value of the final model on the other.

The overall approach in mathematical modeling is illustrated in Figure 2.1. Needless to say, each of these steps involves more detailed work and, as mentioned earlier, will include feedback, iteration, and refinement. In the following sections, a logical approach to the model development process is presented, identifying the various tasks involved in each of the steps. It is not the intention here to propose this as the standard procedure for every modeler to follow in every situation; however, most of the important and crucial tasks are identified and included in the proposed procedure.

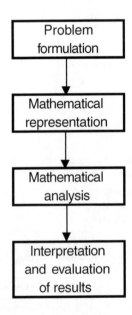

**Figure 2.1** Overall approach to mathematical modeling.

## 2.2.1 PROBLEM FORMULATION

As in any other field of scientific study, formulation of the problem is the first step in the mathematical model development process. This step involves the following tasks:

*Task 1:* establishing the goal of the modeling effort. Modeling projects may be launched for various reasons, such as those pointed out in Chapter 1. The scope of the modeling effort will be dictated by the objective(s) and the expectation(s). Because the premise of the effort is for the model to be simpler than the real system and at the same time be similar to it, one of the objectives should be to establish the extent of correlation expected between model predictions and performance of the real system, which is often referred to as performance criteria. This is highly system specific and will also depend on the available resources such as the current knowledge about the system and the tools available for completing the modeling process.

It should also be noted that the same system might require different types of models depending on the goal(s). For example, consider a lake into which a pollutant is being discharged, where it undergoes a decay process at a rate estimated from empirical methods. If it is desired to determine the long-term concentration of the pollutant in the lake or to do a sensitivity study on the estimated decay rate, a simple static model will suffice. On the other hand, if it is desired to trace the temporal concentration profile due to a partial shutdown of the discharge into the lake, a dynamic model would be required. If toxicity of the pollutant is a key issue and, hence, if peak concentrations due to inflow fluctuations are to be predicted, then a probabilistic approach may have to be adapted.

Another consideration at this point would be to evaluate other preexisting "canned" models relating to the project at hand. They are advantageous because many would have been validated and/or accepted by regulators. Often, such models may not be applicable to the current problem with or without minor modifications due to the underlying assumptions about the system, the contaminants, the processes, the interactions, and other concerns. However, they can be valuable in guiding the modeler in developing a new model from the basics.

*Task 2:* characterizing the system. In terms of the definitions presented earlier, characterizing the system implies identifying and defining the system, its boundaries, and the significant and relevant variables and parameters. The modeler should be able to establish how, when, where, and at what rate the system interacts with its surroundings; namely, provide data about the inflow rates and the outflow rates. Processes and reactions occurring inside the system boundary should also be identified and quantified.

Often, creating a schematic, graphic, or pictographic model of the system (a two-dimensional model) to visualize and identify the boundary and the system-surroundings interactions can be a valuable aid in developing the mathematical model. These aids may be called conceptual models and can include the model variables, such as the directions and the rates of flows crossing the boundary, and parameters such as reaction and process rates inside the system. Jorgensen (1994) presented a comprehensive summary of 10 different types of such tools, giving examples and summarizing their characteristics, advantages, and disadvantages.

Some of the recent software packages (to be illustrated later) have taken this idea to new heights by devising the diagrams to be "live." For example, in a simple block diagram, the boxes with interconnecting arrows can be encoded to act as reservoirs, with built-in mass balance equations. With the passage of time, these blocks can "execute" the mass balance equation and can even animate the amount of material inside the box as a function of time. Examples of such diagrams can be found throughout this book.

Another very useful and important part of this task is to prepare a list of all of the variables along with their fundamental dimensions (i.e., $M, L, T$) and the corresponding system of units to be used in the project. This can help in checking the consistency among variables and among equations, in troubleshooting, and in determining the appropriateness of the results.

*Task 3:* simplifying and idealizing the system. Based on the goals of the modeling effort, the system characteristics, and available resources, appropriate assumptions and approximations have to be made to simplify the system, making it amenable to modeling within the available resources. Again, the primary goal is to be able to replicate or reproduce significant behaviors of the real system. This involves much experience and professional judgement and an overall appreciation of the efforts involved in modeling from start to finish.

For example, if the processes taking place in the system can be approximated as first-order processes, the resulting equations and the solution procedures can be considerably simpler. Similar benefits can be gained by making assumptions: using average values instead of time-dependent values, using estimated values rather than measured ones, using analytical approaches rather than numerical or probability-based analysis, considering equilibrium vs. nonequilibrium conditions, and using linear vs. nonlinear processes.

## 2.2.2 MATHEMATICAL REPRESENTATION

This is the most crucial step in the process, requiring in-depth subject matter expertise. This step involves the following tasks:

*Task 1:* identifying fundamental theories. Fundamental theories and principles that are known to be applicable to the system and that can help achieve the goal have to be identified. If they are lacking, ad hoc or empirical relationships may have to be included. Examples of fundamental theories and principles include stoichiometry, conservation of mass, reaction theory, reactor theory, and transport mechanisms. A review of theories of environmental processes is included in Chapter 4. A review of engineered environmental systems is presented in Chapter 5. And, a review of natural environmental systems is presented in Chapter 6.

*Task 2:* deriving relationships. The next step is to apply and integrate the theories and principles to derive relationships between the variables of significance and relevance. This essentially transforms the real system into a mathematical representation. Several examples of derivations are included in the following chapters.

*Task 3:* standardizing relationships. Once the relationships are derived, the next step is to reduce them to standard mathematical forms to take advantage of existing mathematical analyses for the standard mathematical formulations. This is normally done through standard mathematical manipulations, such as simplifying, transforming, normalizing, or forming dimensionless groups.

The advantage of standardizing has been referred to earlier in Chapter 1 with Equations (1.1) and (1.2) as examples. Once the calculus that applies to the system has been identified, the analysis then follows rather routine procedures. (The term calculus is used here in the most classical sense, denoting formal structure of axioms, theorems, and procedures.) Such a calculus allows deductions about any situation that satisfies the axioms. Or, alternatively, if a model fulfills the axioms of a calculus, then the calculus can be used to predict or optimize the performance of the model. Mathematicians have formalized several calculi, the most commonly used being differential and integral. A review of the calculi commonly used in environmental systems is included in Chapter 3, with examples throughout this book.

## 2.2.3 MATHEMATICAL ANALYSIS

The next step of analysis involves application of standard mathematical techniques and procedures to "solve" the model to obtain the desired results. The convenience of the mathematical representation is that the resulting model can be analyzed on its own, completely disregarding the real system, temporarily. The analysis is done according to the rules of mathematics, and the system has nothing to do with that process. (In fact, any analyst can perform this task—subject matter expertise is not required.)

The type of analysis to be used will be dictated by the relationships derived in the previous step. Generalized analytical techniques can fall into algebraic, differential, or numerical categories. A review of selected analytical techniques commonly used in modeling of environmental systems is included in Chapter 3.

## 2.2.4 INTERPRETATION AND EVALUATION OF RESULTS

It is during this step that the iteration and model refinement process is carried out. During the iterative process, performance of the model is compared against the real system to ensure that the objectives are satisfactorily met. This process consists of two main tasks—calibration and validation.

*Task 1:* calibrating the model. Even if the fundamental theorems and principles used to build the model described the system truthfully, its performance might deviate from the real system because of the inherent assumptions and simplifications made in Task 3, Section 2.2.1 and the assumptions made in the mathematical analysis. These deviations can be minimized by calibrating the model to more closely match the real system.

In the calibration process, previously observed data from the real system are used as a "training" set. The model is run repeatedly, adjusting the model parameters by trial and error (within reasonable ranges) until its predictions under similar conditions match the training data set as per the goals and performance criteria established in Section 2.2.1. If not for computer-based modeling, this process could be laborious and frustrating, especially if the model includes several parameters.

An efficient way to calibrate a model is to perform preliminary sensitivity analysis on model outputs to each parameter, one by one. This can identify the parameters that are most sensitive, so that time and other resources can be allocated to those parameters in the calibration process. Some modern computer modeling software packages have sensitivity analysis as a built-in feature, which can further accelerate this step.

If the model cannot be calibrated to be within acceptable limits, the modeler should backtrack and reevaluate the system characterization and/or the model formulation steps. Fundamental theorems and principles as well as the model formulation and their applicability to the system may have to be reexamined, assumptions may have to be checked, and variables may have to be evaluated and modified, if necessary. This iterative exercise is critical in establishing the utility value of the model and the validity of its applications, such as in making predictions for the future.

*Task 2:* validating the model. Unless a model is well calibrated and validated, its acceptability will remain limited and questionable. There are no standard benchmarks for demonstrating the validity of models, because models have to be linked to the systems that they are designed to represent.

Preliminary, informal validation of model performance can be conducted relatively easily and cost-effectively. One way of checking overall performance is to ensure that mass balance is maintained through each of the model runs. Another approach is to set some of the parameters so that a closed algebraic solution could be obtained by hand calculation; then, the model outputs can be compared against the hand calculations for consistency. For example, by setting the reaction rate constant of a contaminant to zero, the model may be easier to solve algebraically and the output may be more easily compared with the case of a conservative substance, which may be readily obtained. Other informal validation tests can include running the model under a wide range of parameters, input variables, boundary conditions, and initial values and then plotting the model outputs as a function of space or time for visual interpretation and comparison with intuition, expectations, or similar case studies.

For formal validation, a "testing" data set from the real system, either historic or generated expressly for validating the model, can be used as a benchmark. The calibrated model is run under conditions similar to those of the testing set, and the results are compared against the testing set. A model can be considered valid if the agreement between the two under various conditions meets the goal and performance criteria set forth in Section 2.2.1. An important point to note is that the testing set should be completely independent of, and different from, the training set.

A common practice used to demonstrate validity is to generate a parity plot of predicted vs. observed data with associated statistics such as goodness of fit. Another method is to compare the plots of predicted values and observed data as a function of distance (in spatially varying systems) or of time (in temporally varying systems) and analyze the deviations. For example, the number of turning points in the plots and maxima and/or minima of the plots and the locations or times at which they occur in the two plots can be used as comparison criteria. Or, overall estimates of absolute error or relative error over a range of distance or time may be quantified and used as validation criterion.

Murthy et al. (1990) have suggested an index $J$ to quantify overall error in dynamic, deterministic models relative to the real system under the same input $u(t)$ over a period of time $T$. They suggest using the absolute error or the relative error to determine $J$, calculated as follows:

$$J = \int_0^T e(t)^T e(t)\,dt \quad \text{or} \quad J = \int_0^T \tilde{e}(t)^T \tilde{e}(t)\,dt$$

where $e(t) = y_s(t) - y_m(t)$ 　or 　$\tilde{e}(t) = \dfrac{e(t)}{y_s(t)}$

　　$y_s(t)$ = output observed from the real system as a function of time, $t$
　　$y_m(t)$ = output predicted by the model as a function of time, $t$

## 2.2.5 SUMMARY OF THE MATHEMATICAL MODEL DEVELOPMENT PROCESS

In Chapter 1, physical modeling, empirical modeling, and mathematical modeling were alluded to as three approaches to modeling. However, as could be gathered from the above, they complement each other and are applied together in practice to complete the modeling task. Empirical models are used to fill in where scientific theories are nonexistent or too complex (e.g., non-linear). Experimental or physical model results are used to develop empirical models and calibrate and validate mathematical models.

The steps and tasks described above are summarized schematically in Figure 2.2. This scheme illustrates the feedback and iterative nature of the process as described earlier. It also shows how the real system and the "abstract" mathematical system interact and how experimentation with the real system and/or physical models is integrated with the modeling process. It is hoped that the above sections accented the science as well as the art in the craft of mathematical model building.

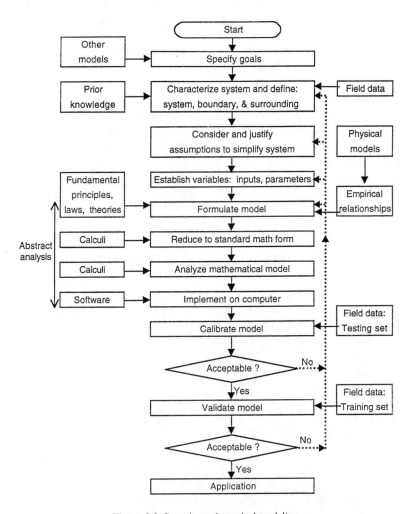

**Figure 2.2** Steps in mathematical modeling.

## 2.3 APPLICATION OF THE STEPS IN MATHEMATICAL MODELING

In this section, the application of the above steps in modeling the spatial variation of a chemical in an advective-dispersive river is detailed. Because the focus of this book is on computer-based mathematical modeling using authoring software tools, the validation and calibration steps are only briefly illustrated. It should not be construed in any way that they are of less importance in the modeling process. Other references must be consulted for further in-depth treatment of these steps.

*Step 1:* Problem Formulation

An industry is proposing to discharge its waste stream containing a toxicant into a nearby river. It is desired to set up discharge permits based on the impact of the discharge on the water quality in the river. A preliminary model has to be developed to predict in-river concentrations of the toxicant under various discharge conditions and river flow rates under steady state conditions.

*Task 1:* establishing a goal. The model has to replicate long-term, spatial variations of the toxicant concentrations in the river. Because the real system per se (river receiving the discharge) is not yet in existence, the desired extent of correlation between the model and real system cannot be established in this case. However, let us pretend that from observations of similar systems, the concentration profile can be expected to have cusp at the point of discharge, with exponential-type decreases on either side. The model should be able to reproduce such a profile. It should enable the users to estimate in-stream concentrations up to 30 miles upstream and up to 50 miles downstream and to determine the maximum discharge that can be allowed without violating certain water quality standards at the above two points. A decision to build a detailed model will be made depending on the results from this preliminary model.

*Task 2:* characterizing the system. The system in this case is the water column, enclosed for the most part by the physical boundaries provided by the riverbed and the banks. An imaginary cross-section at the upstream point of interest, A-A, and another imaginary section at the downstream point of interest, B-B, will complete the boundary as shown in Figure 2.3. The inflow crosses the boundary at A-A, and the outflow crosses at B-B. The origin for measuring distance along the river is taken as the point of discharge.

The most significant variables in this problem can be identified as the flow rate in the river, the concentration of the toxicant in the inflow, the waste input rate, the reaction rate constants for the various processes that the toxicant can undergo within the system, and the length of the river system. Other variables can be the area of flow and the velocity of flow in the river. Some

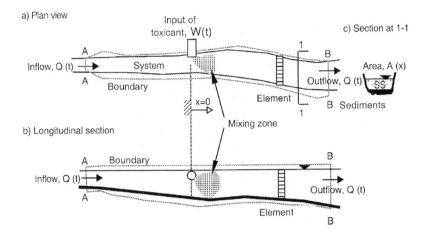

**Figure 2.3** Schematic of real system.

of the environmental processes that the toxicant can undergo within the system, such as adsorption, desorption, volatilization, hydrolysis, photolysis, biodegradation, and biouptake, are illustrated in Figure 2.4.

*Task 3:* simplifying the system. In general, the interactions between the toxicant and the suspended solids and sediments can be under equilibrium, nonequilibrium, linear, or nonlinear conditions. The dissolved form of the toxicant can undergo a variety of processes or reactions that can be of first or higher order. The system parameters, such as area of cross-section and the flow rate of the river, etc., can vary spatially or temporally.

Considering the project goals agreed upon and the physical, chemical, and biological properties of the toxicant and the system, it may be reasonable to make the following simplifying assumptions for the preliminary model:

- Instantaneous mixing in the $z$- and $y$-directions occurs at the point of discharge; thus, the problem is reduced to a one-dimensional analysis.
- Only the dissolved concentration of the toxicant is of significance and relevance; thus, interactions with sediments through suspended solids are negligible.
- All reactions undergone by the toxicant are of first order; thus, all the individual reaction rate constants, $k_i$, can be lumped to an overall rate constant, $K = \Sigma k_i$.
- The river is of fixed prismatic section; thus, the area of flow is time-invariant and space-invariant within the system boundary.
- The dispersion coefficient is significant only in the $x$-direction and remains constant within the system boundary.
- The flow rate is an average constant value, but can be changed, viz., a parameter.

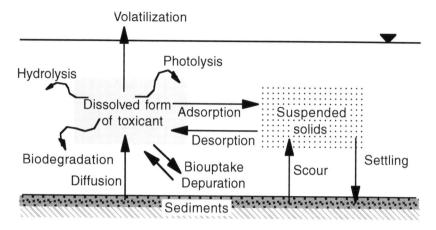

**Figure 2.4** Environmental processes acting upon dissolved toxicant.

- The river flow and waste input are the only inflows into the system, and the river flow is the only outflow from the system.
- The volumetric flow of the waste stream is negligible compared to the river flow.

With these assumptions, the system is reduced to a simplified form as shown in Figure 2.5. The variables of significance and relevance are listed in Table 2.1 along with their respective symbols, dimensions, and common units.

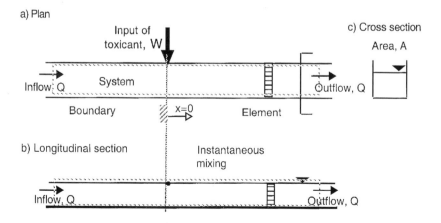

**Figure 2.5** Schematic of the simplified system.

Table 2.1 **Variables, Symbols, Dimensions, and Units**

| Variable | Symbol | Dimension | Unit |
|---|---|---|---|
| River flow rate | $Q$ | $L^3T^3$ | cfs |
| Concentration of toxicant | $C$ | $ML^{-3}$ | mg/L |
| First-order reaction rate constant | $k$ | $T^{-1}$ | 1/day |
| Area of flow | $A$ | $L^2$ | sq ft |
| Velocity of flow | $U$ | $LT^{-1}$ | ft/s |
| Dispersion coefficient | $E$ | $L^2T^{-1}$ | sq miles/day |
| Length in direction of flow | $x$ | $L$ | miles |
| Time | $t$ | $T$ | day |

***Step 2:*** Mathematical Representation

*Task 1:* identifying fundamental theories. Having simplified the system, the appropriate fundamental theorems, laws, and principles necessary to build the model to achieve the goal can be identified as follows:

- Toxicant mass balance based on conservation of mass:

$$\begin{array}{l}\text{Rate of change} \\ \text{of mass of} \\ \text{toxicant inside} \\ \text{system}\end{array} = \begin{array}{l}\text{Net rate of} \\ \text{advective} \\ \text{inflow of} \\ \text{toxicant}\end{array} + \begin{array}{l}\text{Net rate of} \\ \text{diffusive} \\ \text{inflow of} \\ \text{toxicant}\end{array} - \begin{array}{l}\text{Net rate of} \\ \text{loss of} \\ \text{toxicant due} \\ \text{to reactions}\end{array} \qquad (2.1)$$

- Advective mass flow rate based on continuity:

$$= [\text{river flow rate, } Q] \times [\text{concentration of toxicant, } C] \qquad (2.2)$$

- Dispersive mass flow rate based on Fick's Law:

$$\begin{aligned} &= [\text{dispersion coefficient, } E] \times [\text{area of flow, } A] \\ &\quad \times [\text{concentration gradient, } \partial C/\partial x] \end{aligned} \qquad (2.3)$$

- Mass loss due to reaction based on first-order reaction kinetics:

$$= [\text{reaction rate constant, } k] \times [\text{volume, } V] \times [\text{concentration, } C] \qquad (2.4)$$

If other processes, such as interactions with suspended solids, bioaccumulation, volatilization, etc., are to be included, then appropriate principles and relationships have to be included in this listing.

*Task 2:* deriving relationships. The fundamentals identified in Task 2 can now be combined to derive an expression for the output(s). This involves standard mathematical manipulations such as simplification, substitution, and rearrangement.

In this example, a distributed model is appropriate because of spatial variations of the output. The calculi of "differential calculus" is applied to derive the governing partial differential equation. The fundamentals identified above are applied to a small element of length, $\Delta x$, within the system over an infinitesimal time interval, $\Delta t$, as illustrated in Figure 2.6.

The components of the mass balance equation for the toxicant can be expressed in terms of the symbols as follows:

$$\text{Rate of change of mass of toxicant within the element} = \frac{\partial(VC)}{\partial t} \quad (2.5)$$

$$\text{Advective inflow rate into element} = QC \quad (2.6)$$

$$\text{Advective outflow rate from element} = QC + \frac{\partial(QC)}{\partial x}\Delta x \quad (2.7)$$

$$\therefore \text{ Net advective inflow} = QC - \left\{ QC + \frac{\partial(QC)}{\partial x}\Delta x \right\} = -\frac{\partial(QC)}{\partial x}\Delta x \quad (2.8)$$
from element

$$\text{Diffusive inflow rate into element} = -EA\frac{\partial C}{\partial x} \quad (2.9)$$

$$\text{Diffusive outflow rate from element} = -EA\frac{\partial C}{\partial x} + \frac{\partial}{\partial x}\left(-EA\frac{\partial C}{\partial x}\right)\Delta x \quad (2.10)$$

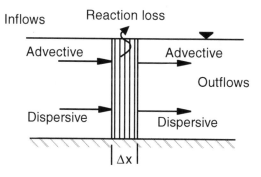

**Figure 2.6** Detail of element.

$$\therefore \text{Net diffusive inflow into element} = \left\{-EA\frac{\partial C}{\partial x}\right\} - \left\{-EA\frac{\partial C}{\partial x} + \frac{\partial}{\partial x}\left(-EA\frac{\partial C}{\partial x}\right)\Delta x\right\}$$

$$= EA\frac{\partial}{\partial x}\left(\frac{\partial C}{\partial x}\right)\Delta x \tag{2.11}$$

Net rate of loss of toxicant due to reactions
within the element $= k(A\Delta x)C$ \hfill (2.12)

Combining the above expressions now completes the mass balance equation:

$$\frac{\partial(VC)}{\partial t} = -\frac{\partial(QC)}{\partial x}\Delta x + \frac{\partial}{\partial x}\left(EA\frac{\partial C}{\partial x}\right)\Delta x - k(A\Delta x)C \tag{2.13}$$

The solution to the above partial differential equation will yield the output $C$ as a function of time, $t$, and distance, $x$.

Based on the goals and the assumptions, the model has to describe steady state conditions, and temporal variations are not required; hence, the partial derivative term on the left-hand side (LHS) of the above equation equals zero, and the equation becomes an ordinary differential equation. Further, $Q$ is a constant, but a parameter; $E$ and $A$ are assumed constant within the boundary; thus, the above equation reduces to the following:

$$0 = -Q\frac{dC}{dx}\Delta x + EA\frac{d^2C}{dx^2}\Delta x - k(A\Delta x)C \tag{2.14}$$

which on division by $(A\Delta x)$ simplifies further to:

$$0 = -\left(\frac{Q}{A}\right)\frac{dC}{dx} + E\frac{d^2C}{dx^2} - kC \tag{2.15}$$

*Task 3:* standardizing relationships. The relationship(s) derived from the fundamental theories and principles can now be translated into standard mathematical forms for further manipulation, analysis, and solving. Mathematical handbooks containing solutions to standard formulations have to be referred to in order to identify the ones matching the equations derived.

In this example, the final equation can be recognized as a second-order, ordinary, homogeneous, linear differential equation, of the standard form:

$$0 = ay'' + by' + cy \tag{2.16}$$

whose solution can be found from handbooks as:

$$y = Me^{gx} + Ne^{jx} \tag{2.17}$$

where $g$ and $j$ are, in turn, the positive and negative values of $-b \pm \sqrt{b^2 - 4ac}/2a$, and $M$ and $N$ are constants to be found from two boundary conditions.

***Step 3:*** Mathematical Analysis

Comparing the coefficients of the final equation derived in our example with those of the standard equation, $a = E$, $b = -(Q/A)$, and $c = -K$. Recognizing that $Q/A = U$, the river velocity, $b = -U$. Thus, $g$ and $j$ are first found as follows:

$$g, j = \frac{U \pm \sqrt{U^2 + 4Ek}}{2E} = \frac{U}{2E}(1 \pm \alpha) \tag{2.18}$$

$$\text{where, } \alpha = \sqrt{1 + \frac{4Ek}{U^2}} \tag{2.19}$$

Now, the constants $M$ and $N$ have to be determined using appropriate boundary conditions for this particular situation. (This is another reason for establishing the system boundary in advance.) It is convenient in this case to consider the reaches upstream and downstream of the discharge point as two separate regions to determine the boundary conditions (BCs). Assuming the toxicant concentration far upstream and far downstream from the discharge point will approach near-zero values, the following boundary BCs can be used to find $M$ and $N$ and, hence, the final solution:

Upstream reach: $x \le 0$
        BC 1: $x = 0$, $C = C_0$
        BC 2: $x = -\infty$, $C = 0$
        $M = C_0$
        $N = 0$

and,

        $C = C_0 e^{gx}$

Downstream reach: $x \ge 0$
        BC 1: $x = 0$, $C = C_0$
        BC 2: $x = +\infty$, $C = 0$
        $N = C_0$
        $M = 0$

and,

        $C = C_0 e^{jx}$

It remains to determine $C_0$ to complete the solution. A mass balance across an infinitesimal element at the discharge point can now yield the value for $C_0$

in terms of the model parameters already defined. This element can be treated as a subsystem at the discharge point as shown in Figure 2.7.

The mass inflow, outflow, and reaction loss for this element are as follows:

$$\text{Mass inflow into element} = QC_0 - EA\left(\frac{dC}{dx}\right)_{x=0-\varepsilon} \qquad (2.20)$$

$$\text{Mass outflow from element} = QC_0 - EA\left(\frac{dC}{dx}\right)_{x=0+\varepsilon} \qquad (2.21)$$

$$\text{Reaction loss within element} = kCA\Delta x = 0$$

Hence, the mass balance equation for the element, ignoring the reaction loss in the infinitesimal element, is as follows:

$$QC_0 - EA\left(\frac{dC}{dx}\right)_{x=0-\varepsilon} + W = QC_0 - EA\left(\frac{dC}{dx}\right)_{x=0+\varepsilon} \qquad (2.22)$$

Now, substituting from the previous expressions for $C$,

$$-EA\left(\frac{d[C_0 e^{gx}]}{dx}\right)_{x=0-\varepsilon} + W = -EA\left(\frac{[C_0 e^{jx}]}{dx}\right)_{x=0+\varepsilon} \qquad (2.23)$$

and simplifying,

$$-EAC_0 g + W = -EAC_0 j \qquad (2.24)$$

or,

$$C_0 = \frac{W}{EA(g-j)} \qquad (2.25)$$

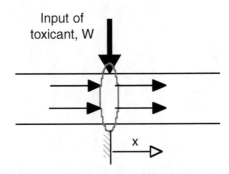

**Figure 2.7** Element at point of discharge.

On substituting for $g$ and $j$ from the above, $C_0 = W/\alpha Q$, where $W$ is the mass rate of discharge into the river. The final model for this problem is, therefore:

$$C = \frac{W}{\alpha Q} e^{gx} \quad \text{for } x \leq 0 \qquad (2.26)$$

$$C = \frac{W}{\alpha Q} e^{jx} \quad \text{for } x \geq 0 \qquad (2.27)$$

***Step 4:*** Interpretation of Results

The calibration and validation of the model will be highly problem-specific. Initial interpretations can include simulations, sensitivity analysis, and comparison with other similar systems. As a first step in this case, the model can be run with typical parameters and known inputs. The output can be used to corroborate the performance of the model against intuition, past experience, or literature results to verify that the model outputs generally follow the observed profiles and are within reasonable ranges. Because the result in this case is a function of $x$ rather than a numerical value, it may be useful to plot the variation of $C$ as a function of $x$ to check if the model is reflecting the spatial concentration profile.

As an example of calibration and validation, let us use an artificial data set (adapted from Thomann and Mueller, 1987):

| Distance (mi) | −15 | −10 | −5 | 0 | 5 | 10 | 15 |
|---|---|---|---|---|---|---|---|
| Concentration (mg/L) | 0.43 | 1.22 | 3.50 | 10.00 | 4.49 | 2.02 | 0.91 |

This data set had been collected on an estuary flowing at 100 cfs with an average cross-sectional area of $10 \times 10^5$ sq ft, receiving a waste input of 372,000 lbs/day at 0 miles. The decay rate of the waste material was measured to be 0.1 $day^{-1}$. The dispersion for this estuary was estimated to be in the range of 2–5 sq miles/day.

As a first step, an appropriate value for the dispersion coefficient, $E$, has to be established. The data from the upstream portion of the estuary are used to calibrate the model to fit the four data points by running the model with various values of $E$ ranging from 2 to 5 sq miles/day. Given the value of $E$, the remaining three data points may be used to validate the model. These considerations are illustrated in Figure 2.8, from which the value of $E$ can be estimated to be 3 sq miles/day. With this value for $E$, the model is capable of fitting the upstream data points as well as predicting the downstream data points. While acknowledging that the validation and calibration exercise in this example is somewhat academic, the very same procedure is used in practice to calibrate and validate mathematical models.

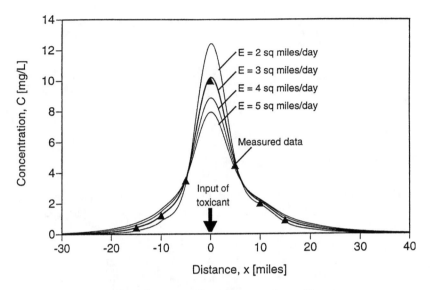

**Figure 2.8** Model predictions vs. measured data.

The value of the simplifying assumptions made at the beginning of the analysis is in the reduction of the governing equation to a form amenable to a closed solution. Such an analysis provides highly useful insight and acts as a stepping stone for the modeler to gradually add pertinent details and refinements to build more realistic models. Needless to say, as the models become more detailed and realistic, analytical methods of solving the governing equations can be extremely complicated or impossible. In such cases, using computer-based, numerical approaches is the only way to complete the modeling process. In the next chapter, some of the common analytical and numerical approaches used in environmental modeling efforts will be reviewed.

## CHAPTER 3

# Primer on Mathematics

## CHAPTER PREVIEW

*This chapter contains a review of mathematical methods and tools as they apply to environmental modeling. It is assumed that the readers have taken a formal sequence of college-level course work in mathematics leading up to partial differential equations. In the first part of this chapter, reviews of different types of mathematical formulations are summarized. Analytical and numerical procedures for solving them are outlined. Then, ways to implement some of the more common procedures in the computer environment are demonstrated.*

## 3.1 MATHEMATICAL FORMULATIONS

IN the previous chapter, several steps and tasks involved in the model development process were identified. It goes without saying that a clear understanding of mathematical formulations and analyses is a necessary prerequisite in this process. A strong mathematical foundation is required to transmute subject matter knowledge into mathematical forms such as functions, expressions, and equations. Knowledge of analytical procedures in mathematical calculi such as simplifying, transforming, and solving, is essential to select and develop the appropriate computational procedures for computer implementation. The selection of an appropriate computer software package to complete the model also requires a good understanding of the mathematics underlying the model. As pointed out in Chapter 1, different formulations can be developed to describe the same system; hence, the ability to choose the optimal one that can meet the goals requires a strong mathematical foundation.

The advantage of reducing the formulations to standard mathematical forms has been pointed out before. For modeling purposes, a wide variety of environmental systems can be categorized as deterministic with continuous variables. Deterministic systems can be described either by static or dynamic formulations. This chapter will, therefore, focus on the mathematical calculi for continuous, deterministic, static, and dynamic systems with up to four independent variables. Brief discussions of how these deterministic models can be adapted for probability systems will be illustrated in selected cases in later chapters. In the following sections, selected standard mathematical formulations commonly encountered in modeling environmental systems are reviewed. In a broad sense, these formulations can be classified as either static or dynamic.

## 3.1.1 STATIC FORMULATIONS

Static models are often built of algebraic equations. The general standard form of the algebraic equation in static formulations is as follows:

$$G(x,y,\theta) = 0 \tag{3.1}$$

where $G$ is a vector function, $x$ and $y$ are vector variables, and $\theta$ is a set of parameters. In the context of a model, $x$ can correspond to the inputs, $y$ the outputs, and $\theta$ the system parameters. If $x$ and $y$ are linear in $G$, the model is called linear, otherwise, it is nonlinear.

## 3.1.2 DYNAMIC FORMULATIONS

Dynamic systems with continuous variables are normally described by differential equations. Any equation containing one or more derivatives is called a differential equation. When the number of independent variables in a differential equation is not more than 1, the equation is called an ordinary differential equation (ODE), otherwise, it is called a partial differential equations (PDE).

The general standard form of an ODE is as follows:

$$\alpha(t)\frac{d^n z(t)}{dt^n} = G\left\{ z(t), \frac{dz(t)}{dt}, \frac{d^2 z(t)}{dt^2}, \ldots \frac{d^{n-1} z(t)}{dt^{n-1}}, u(t), \frac{du(t)}{dt}, \right.$$

$$\left. \frac{d^2 u(t)}{dt^2}, \ldots, \theta(t) \right\} \tag{3.2}$$

where $u(t)$ is a known function, $\alpha(t)$ and $\theta(t)$ are parameters, and $z(t)$ is the dependent variable. In the context of a model, $t$ can correspond to time, $u$ to the input, $z$ the output, and $\alpha(t)$ and $\theta(t)$ the system parameters.

An ODE is ranked as of order $n$ if the highest derivative of the dependent variable is of order $n$. When $\alpha(t)$ is nonzero, i.e., the equation is nonsingular, it can be simplified by dividing throughout by $\alpha(t)$. Often, in many environmental systems, $\theta(t)$ does not change with $t$. Further, if $G$ is linear in $z(t)$, $u(t)$, and their derivatives, then the equation is linear, and the principle of superpositioning can be applied.

The general standard form of a PDE with two independent variables $x$ and $t$ is as follows:

$$G[u, u_x, u_t, u_{xx}, u_{xt}, u_{tt}, \theta(x,t), f(x,t)] = 0 \qquad (3.3)$$

where

$$u_x = \frac{\partial u}{\partial x}; \ u_t = \frac{\partial u}{\partial t}; \ u_{xx} = \frac{\partial^2 u}{\partial t^2}; \ u_{xt} = \frac{\partial^2 u}{\partial t^2}; \ u_{tt} = \frac{\partial^2 u}{\partial t^2}, \ \theta(x,t)$$

is a parameter and $f(x,t)$ is a known function. In the context of a model, $x$ can correspond to a spatial coordinate, $t$ to time, $f(x,t)$ to the input, $\theta(x,t)$ to the system parameter, and $u(x,t)$ to the system outputs. The formulation should also define the problem domain, i.e., ranges for $x$ and $t$.

## 3.2 MATHEMATICAL ANALYSIS

Some of the formulations identified above are tractable to an *analytical* method of analysis, while many require a *computational* (also referred to as *numerical*) method of analysis. Both methods of analysis can form the basis of computer-based mathematical modeling.

### 3.2.1 ANALYTICAL METHODS

In analytical methods, the solution to a formulation is found as an *expression* consisting of the parameters and the independent variables in terms of the symbols. Sometimes this method of solution is referred to as parameterized solutions. The solution can be exact or approximate. Only for a limited class of formulations is it possible to find an exact analytical solution. An approximate solution has to be sought in other cases, such as in the models for large environmental systems.

For example, consider Equation (3.1), where $y$ is the unknown, $x$ is a known variable, and $\theta$ is a parameter. The solution for $y$ might or might not exist; if it does, it might not be unique. If Equation (3.1) could be rearranged to the form

$$A(x,\theta)y + B(x,\theta) = 0 \qquad (3.4)$$

a solution can be found by inverting the matrix $A(x,\theta)$ if and only if $A(x,\theta)$ is a nonsingular $n \times n$ matrix. If $y$ appears nonlinearly, multiple solutions may be possible, and a computational method of analysis would be necessary to find them.

## 3.2.2 COMPUTATIONAL METHODS

In this method, the solution is found numerically, often with the aid of a computer. The solutions in this case are only approximate, and *numeric*. Therefore, the formulation should include numeric values for the model parameters and variables, whereas symbolic representations will suffice in the case of the analytical method of analysis. The advantage of the computational method of analysis is that it can be applied to a much wider class of mathematical formulations, particularly for complex systems.

It is not within the scope of this book to identify all the standard mathematical calculi and procedures for analyses, rather, some specific examples of analyses pertaining to typical environmental systems are illustrated. The intent of this illustration is for the readers to be able to adapt them for implementation in the computer environment.

## 3.3 EXAMPLES OF ANALYTICAL AND COMPUTATIONAL METHODS

The governing equations in environmental models may be reduced to simple algebraic equations (e.g., steady state concentration of a contaminant in a completely mixed lake), systems of simultaneous linear equations (e.g., steady state concentrations in completely mixed lakes in series), ODEs (e.g., transient concentration in a well-mixed lake), systems of ODEs (e.g., biomass growth, substrate consumption, and oxygen level in a completely mixed lake), or PDEs (e.g., contaminant transport in a stratified lake under transient loads). In this section, common algorithms for solving these types of formulations are outlined. Many of these algorithms can be implemented in spreadsheet programs with minimal syntax or programming. Many standard algorithms are included as preprogrammed libraries in other software packages discussed in this book. Some examples of implementations are included in this chapter to illustrate the general approach, and more detailed ones for specific problems can be found in Chapters 8 and 9.

### 3.3.1 ALGEBRAIC EQUATIONS

#### 3.3.1.1 Classifications of Algebraic Equations

The most general form of the algebraic equation was given in Equation (3.1), and a solvable form was provided in Equation (3.4). Considering a

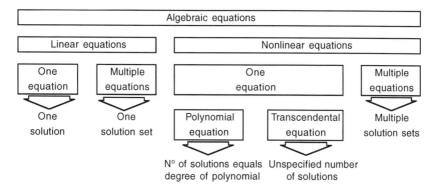

**Figure 3.1** Classification of algebraic equations.

simpler form of $G$, such as $f(x) = 0$, its solution or "root" is the value of the independent variable, $x$, that when substituted into $f(x)$ will make it equal to zero. Methods to find those roots for such equations depend on the number of equations and the type of equations to be solved. Classification of algebraic equations is shown in Figure 3.1 to help in this selection process.

### 3.3.1.2 Single, Linear Equations

Analytical methods of elementary algebra for solving single, linear equations for one unknown are rather straightforward and are not discussed further.

### 3.3.1.3 Set of Linear Equations

Simultaneous linear equations are frequently encountered in environmental modeling. Typical examples include chemical speciation calculations and numerical solution of partial differential equations. A general form of a set of $m$ linear equations with $n$ unknowns is $\mathbf{A}x = \mathbf{B}$, where $\mathbf{A}$ is a given $m \times n$ matrix, and $\mathbf{B}$ is a given vector. The solution is given by $x = \mathbf{A}^{-1}\mathbf{B}$. This equation in general will have a unique solution only if $m = n$; if $m < n$, infinitely many solutions may be possible; and if $m > n$ no solutions are possible. An example of a set of linear simultaneous equations is as follows:

$$a_{11}x_1 + a_{12}x_2 + a_{13}x_3 = b_1$$

$$a_{21}x_1 + a_{22}x_2 + a_{23}x_3 = b_2$$

$$a_{31}x_1 + a_{32}x_2 + a_{33}x_3 = b_3$$

or, in matrix form

$$\begin{bmatrix} a_{11} & a_{12} & a_{13} \\ a_{21} & a_{22} & a_{23} \\ a_{31} & a_{32} & a_{33} \end{bmatrix} \begin{bmatrix} x_1 \\ x_2 \\ x_3 \end{bmatrix} = \begin{bmatrix} b_1 \\ b_2 \\ b_3 \end{bmatrix}$$

where $a_{ij}$ are the coefficients, $b$'s are constants, and $x$'s are the unknowns.

The Gauss-Seidel iterative method is a convenient computational method used for solving such a set of equations. The algorithm first solves the first equation for $x_1$, the second for $x_2$, and the third for $x_3$.

$$x_1 = \frac{b_1 - a_{12}x_2 - a_{13}x_3}{a_{11}} \tag{3.5a}$$

$$x_2 = \frac{b_2 - a_{21}x_1 - a_{23}x_3}{a_{22}} \tag{3.5b}$$

$$x_3 = \frac{b_3 - a_{31}x_1 - a_{32}x_2}{a_{33}} \tag{3.5c}$$

The process is iterated with resubstitution, until a desired degree of convergence is reached using Equations (3.5). This algorithm is illustrated in the next example.

### Worked Example 3.1

Solve the following simultaneous equations:

$$4\ X1 + 6\ X2 + 2\ X3 = 11$$
$$2\ X1 + 6\ X2 + \quad X3 = 21$$
$$3\ X1 + 2\ X2 + 5\ X3 = 75$$

### Solution

Figure 3.2 shows an Excel®[3] implementation for solving the three equations, which are entered in rows 2 to 4. Note how the coefficients are entered into separate cells so that they can be referred to by cell reference. The Gauss-Seidel algorithm is entered into rows 6 to 8, in column N, to estimate $X1$, $X2$, and $X3$, respectively.

For illustration, the algorithm is expressed in Excel® language in column L, against each $X$ row. Notice that the formula in cell L6 refers to cell L7, while the formula in cell L7 refers back to L6. This is known as *circular*

---

[3]Excel® is a registered trademark of Microsoft Corporation. All rights reserved.

|   | A | B | C | D | E | F | G | H | I | J | K | L | M | N |
|---|---|---|---|---|---|---|---|---|---|---|---|---|---|---|
| 1 |   |   |   |   |   |   |   |   |   |   |   |   |   |   |
| 2 |   | Equation 1: | 4 | X1 | + | 6 | X2 | + | 2 | X3 | = 11 |   |   |   |
| 3 |   | Equation 2: | 2 | X1 | + | 6 | X2 | + | 1 | X3 | = 21 |   |   |   |
| 4 |   | Equation 3: | 3 | X1 | + | 2 | X2 | + | 5 | X3 | = 75 |   |   |   |
| 5 |   |   |   |   |   |   |   |   |   |   |   |   |   |   |
| 6 |   | **Calculate roots:** |   |   |   |   |   |   | X1 | = (L2-F2*L7-I2*L8)/C2 |   | = -16.38 |   |   |
| 7 |   |   |   |   |   |   |   |   | X2 | = (L3-C3*L6-I3*L8)/F3 |   | = 5.17 |   |   |
| 8 |   |   |   |   |   |   |   |   | X3 | = (L4-C4*L6-F4*L7)/I4 |   | = 22.76 |   |   |
| 9 |   |   |   |   |   |   |   |   |   |   |   |   |   |   |
| 10 |   | **Check b values:** |   |   |   |   |   |   | b1 | = C2*N6+F2*N7+I2*N8 |   | = 11 |   |   |
| 11 |   |   |   |   |   |   |   |   | b2 | = C3*N6+F3*N7+I3*N8 |   | = 21 |   |   |
| 12 |   |   |   |   |   |   |   |   | b3 | = C4*N6+F4*N7+I4*N8 |   | = 75 |   |   |
| 13 |   |   |   |   |   |   |   |   |   |   |   |   |   |   |

**Figure 3.2** Gauss-Siedel algorithm implemented in the Excel® spreadsheet program.

*reference* in Excel® and will cause an error message to be generated. To execute such circular references, the *Iteration* option in the *Calculation* panel under the *Preferences* menu item under the *Tools* menu should be turned on. Once the equations are entered, the spreadsheet can be *Run* to solve the equations iteratively. The results calculated are returned in column N. Finally, the results can be checked by feeding them back into the original equations to ensure that they satisfy them as shown in rows 10 to 12.

Other methods such as the Gauss Elimination method are also available to solve linear simultaneous equations. Most equation solver-based packages feature built-in procedures for solving these equations, requiring minimal programming.

An example of the use of Excel®'s built-in *Solver* utility that can be used to solve a set of equations is presented next. Consider the same set of equations solved in Worked Example 3.1, and let the functions *f, g,* and *h* represent those equations:

$$f \equiv 4\,X1 + 6\,X2 + 2\,X3 - 11$$
$$g \equiv 2\,X1 + 6\,X2 + \ \ X3 - 21$$
$$h \equiv 3\,X1 + 2\,X2 + 5\,X3 - 75$$

Recognizing the fact that the roots of the above equations will make $y = f^2 + g^2 + h^2 = 0$, the problem of finding those roots can be tackled readily by calling the *Solver* routine of Excel® as illustrated in Figure 3.3. Here, $X1$, $X2$, and $X3$ are assigned arbitrary guess values of 1, 2, and 3 in column M. The expression for $y$ is entered into cell J6. The *Solver* routine is selected from the *Tools* menu, and the target cell and the cells to be changed are specified in the *Solver Parameter* dialog box. The routine then will find the values of $X1$, $X2$, and $X3$ that will make $y = 0$.

(a)

| | A | B | C | D | E | F | G | H | I | J | | K | L | M | N | O | P | Q |
|---|---|---|---|---|---|---|---|---|---|---|---|---|---|---|---|---|---|---|
| 2 | | f = | 4 | X1+ | 6 | X2+ | 2 | X3+ | = | 11 | | X1 | = | 1 | | f1 | = | 11 |
| 3 | | g = | 2 | X1+ | 6 | X2+ | 1 | X3+ | = | 21 | | X2 | = | 2 | | f2 | = | -4 |
| 4 | | h = | 3 | X1+ | 2 | X2+ | 5 | X3+ | = | 75 | | X3 | = | 3 | | f3 | = | -53 |
| 6 | | y = | f^2 + g^2 + h^2 | | | | | = | | 2946 | | | | | | | | |

**Solver Parameters**

Set Target Cell: [ $J$6 ]   [Solve]

Equal To:  ○ Max  ○ Min  ● Value of: [0]   [Close]

By Changing Cells:

[ $M$2:$M$4 ]   [Guess]

Subject to the Constraints:   [Options]

[ ]   [Add]

[Change]   [Reset All]

[Delete]   [Help]

(b)

| | A | B | C | D | E | F | G | H | I | J | | K | L | M | N | O | P | Q |
|---|---|---|---|---|---|---|---|---|---|---|---|---|---|---|---|---|---|---|
| 2 | | f = | 4 | X1+ | 6 | X2+ | 2 | X3+ | = | 11 | | X1 | = | -16.4 | | f1 | = | -0 |
| 3 | | g = | 2 | X1+ | 6 | X2+ | 1 | X3+ | = | 21 | | X2 | = | 5.17 | | f2 | = | 0 |
| 4 | | h = | 3 | X1+ | 2 | X2+ | 5 | X3+ | = | 75 | | X3 | = | 22.8 | | f3 | = | -0 |
| 6 | | y = | f^2 + g^2 + h^2 | | | | | = | | 2E-09 | | | | | | | | |

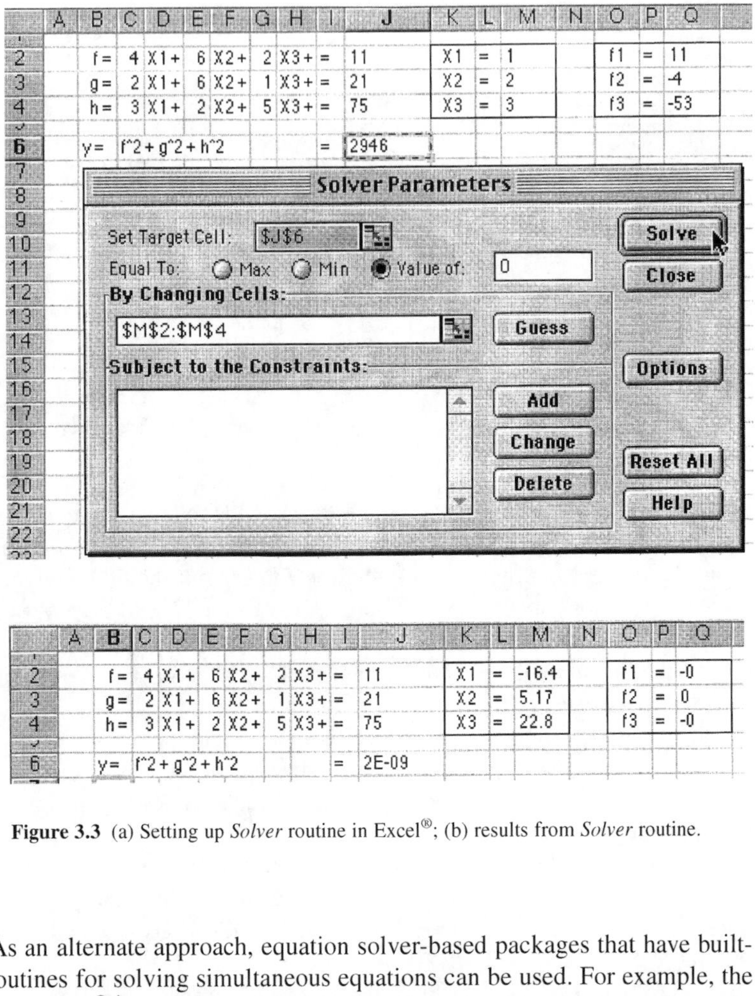

**Figure 3.3** (a) Setting up *Solver* routine in Excel®; (b) results from *Solver* routine.

As an alternate approach, equation solver-based packages that have built-in routines for solving simultaneous equations can be used. For example, the Mathematica® [4] equation solver-based software can be used as shown in Figure 3.4. The coefficients $a_{ij}$ and $b_k$ are first assigned appropriate numerical values. Then, the built-in routine, *Solve,* is called with two lists of arguments. The first list contains all the equations to be solved in symbolic form, and the second list contains the variables for which the equations are to be solved. When executed, the roots of the three equations are returned in line *Out[1]* as $x1 = -16.381$; $x2 = 5.16667$; and $x3 = 22.7619$.

An elegant way to solve a set of linear equations is by following the mathematical calculi of matrix algebra. Another equation solver-type software

[4]Mathematica® is a registered trademark of Wolfram Research, Inc. All rights reserved.

```
In[1]:= a11=4.0; a12=6.0; a13=2.0; b1 = 11.0;
        a21=2.0; a22=6.0; a23=1.0; b2 = 21.0;
        a31=3.0; a32=2.0; a33=5.0; b3 = 75.0;
        Solve[{a11*x1 + a12*x2 + a13*x3 == b1,
        a21*x1 + a22*x2 + a23*x3 == b2, a31*x1
        a32*x2 + a33*x3 ==b3}, {x1, x2, x3}]

Out[1]= {{x1 → -16.381, x2 → 5.16667, x3 → 22.7619}}
```

**Figure 3.4** Using Mathematica® for solving simultaneous equations.

package, MATLAB®[5], allows this to be set up effortlessly. The same set of equations as in the above example is solved in MATLAB® as shown in Figure 3.5. The matrix **a** is first specified with $a_{ij}$, followed by matrix **b** with $b$'s. Then, by entering the command $x = \mathbf{b}/\mathbf{a}$, MATLAB® returns the solution for $x$ with $x1 = -16.3810$; $x2 = 5.1667$; and $x3 = 22.7619$.

### 3.3.1.4 Single, Nonlinear Equations

The next class of algebraic equations is single, nonlinear equations. Solution methods for nonlinear equations are either *direct* or *indirect*. In the direct method, known formulas are applied to standard forms of the equations in a nonrepetitive manner (analytical methods of analysis). A typical example is the standard solution for a second-order polynomial equation, otherwise known as the "quadratic equation": $ax^2 + bx + c = 0$, whose roots are given by $\{-b \pm \sqrt{b^2 - 4ac}\}/2a$.

Such formulas are not readily available or unknown for many types of equations. Hence, indirect methods have to be used in those cases. In the *indirect* method, repeated application of some algorithm is implemented to yield an approximate solution (computational method of analysis). The indirect methods are the ones that are utilized in computer modeling of complex systems.

Nonlinear equations can be either polynomial or transcendental. The computational methods for solving such equations start with a guessed value for the root and follow standard computer algorithms to systematically refine that guess in an iterative manner until the equation is satisfied within acceptable limits. Two simple methods are outlined here.

In the first, known as the *binary method,* two guesses $x_l$ and $x_u$ are made such that they bracket the real root, $x$: $x_l < x$ and $x_u > x$. While this may appear circuitous, as $x$ is not known, $x_l$ and $x_u$ can be found rather easily by taking

---

[5]MATLAB® is a registered trademark of The MathWorks, Inc. All rights reserved.

```
»a=[4,2,3;6,6,2;2,1,5];
»b=[11, 21, 75];
»x=b/a
```

x=

-16.3810  5.1667  22.7619

**Figure 3.5** Using MATLAB® for solving simultaneous equations.

advantage of the fact that the function should change sign within the interval bounded by $x_l$ and $x_u$. Or, in other words, guess $x_l$ and $x_u$ so that:

$$f(x_l,\theta) \bullet f(x_u,\theta) < 0 \tag{3.6}$$

This can be readily achieved by plotting the function. Then, a refined value of the root, $x_r$, can be estimated as = ( $x_l + x_u$)/2. To make the next refined guess, a new bracket is now defined with either $x_l$ and $x_r$ or $x_r$ and $x_u$. Again, a sign change of $f(x)$ is used to decide which range to make the new guess from:

if $f(x_l,\theta) \times f(x_r,\theta) < 0$, the new guess is made between $x_l$ and $x_r$ (3.7a)

if $f(x_r,\theta) \times f(x_u,\theta) < 0$, the new guess is made between $x_r$ and $x_u$ (3.7b)

This process is iterated until the new guess is not significantly different from the previous one; at that point, the root is taken as the value of the last guess.

***Worked Example 3.2***

First-order processes occurring in many environmental systems can be described by the equation: $C = C_0 e^{-kt}$, where $C$ is the concentration of the chemical undergoing the reaction, $C_0$ is its initial concentration, $k$ is the reaction rate constant, and $t$ is the time. Find the time it would take for the concentration to drop from 100 mg/L to 10 mg/L.

*Solution*

Even though the equation can be solved for $t$ algebraically, the binary method is used here to illustrate the method and to compare its performance against the direct algebraic solution. The algebraic solution can be readily seen as $t = 9.21$. The implementation of the binary algorithm in an Excel® spreadsheet is shown in Figure 3.6.

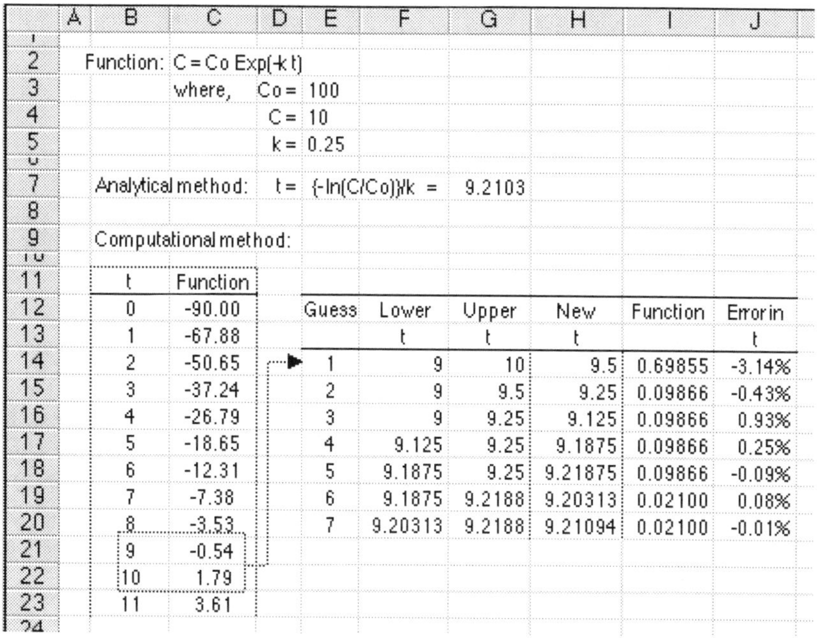

| | t | Function | | Guess | Lower t | Upper t | New t | Function | Error in t |
|---|---|---|---|---|---|---|---|---|---|
| Function: C = Co Exp(-k t) | | | | | | | | | |
| where, Co = 100 | | | | | | | | | |
| C = 10 | | | | | | | | | |
| k = 0.25 | | | | | | | | | |
| Analytical method: | | t = {-ln(C/Co)}/k = | | 9.2103 | | | | | |
| Computational method: | | | | | | | | | |
| | 0 | -90.00 | | | | | | | |
| | 1 | -67.88 | | | | | | | |
| | 2 | -50.65 | | 1 | 9 | 10 | 9.5 | 0.69855 | -3.14% |
| | 3 | -37.24 | | 2 | 9 | 9.5 | 9.25 | 0.09866 | -0.43% |
| | 4 | -26.79 | | 3 | 9 | 9.25 | 9.125 | 0.09866 | 0.93% |
| | 5 | -18.65 | | 4 | 9.125 | 9.25 | 9.1875 | 0.09866 | 0.25% |
| | 6 | -12.31 | | 5 | 9.1875 | 9.25 | 9.21875 | 0.09866 | -0.09% |
| | 7 | -7.38 | | 6 | 9.1875 | 9.2188 | 9.20313 | 0.02100 | 0.08% |
| | 8 | -3.53 | | 7 | 9.20313 | 9.2188 | 9.21094 | 0.02100 | -0.01% |
| | 9 | -0.54 | | | | | | | |
| | 10 | 1.79 | | | | | | | |
| | 11 | 3.61 | | | | | | | |

**Figure 3.6** Implementation of binary method in Excel® spreadsheet.

The first step in the binary method is to guess the lower and upper bounds of the root by calculating the function $C - C_0 e^{-kt}$ for a range of values of $t$ to determine the value of $t$ at which a sign change occurs. It can be readily done in Excel® by entering the function in cell C12 and *filling* it down, with $t$ values set up in column B, again by filling down. The upper and lower bounds are seen to be $t = 9$ and $t = 10$. The following algorithms, based on Equation (3.5), are entered into cells F15 and G15, and filled down to perform the calculations automatically for seven steps, in this case:

Cell F15:    IF((($C$4-Co*EXP(-k*F14))*($C$4-Co*EXP(-k*H14))<0,F14,H14)

Cell G15:    IF((($C$4-Co*EXP(-k*G14))*($C$4-Co*EXP(-k*H14))<0,G14,H14)

As can be seen from Figure 3.6, this procedure quickly converges on the root to a high degree of accuracy. Even though a more elegant spreadsheet can be developed for this application, the intent here is to illustrate the procedure as well as the ease with which a simple model could be developed.

Another method, known as the Newton-Raphson method, requires only one guess to start the iterative process. It is based on Taylor's expansion of the form:

$$f(x_n + h) = f(x_{n+1}) = f(x_n) + hf'(x_n) + \frac{h^2}{2}f'(x_n) + \ldots \qquad (3.8)$$

If $h^2$ and higher terms are ignored, it can be seen that the step from $x_n$ to $x_{n+1}$ moves the function value closer to a root so that $f(x_n + h) = 0$. Then,

$$x_{n+1} = x_n - \frac{f(x_n)}{f(x_n)} \qquad (3.9)$$

Here, the computational process starts with a guess value for $x_n$. Then, using $f(x_n)$ and $f'(x_n)$, a value for $x_{n+1}$ is calculated. If $f(x_{n+1})$ is sufficiently small, the root is taken as $x_{n+1}$; or, a new value for $x_{n+1}$ is calculated using the current value of $x_{n+1}$ for $x_n$ in Equation (3.7), and the process is repeated. A modification to this method is the Secant method, which is preferable when it is difficult to get the derivative $f'(x_n)$ to be used in Equation (3.7). In such cases, the Secant method uses the following approximation for $f'(x_n)$:

$$f'(x_n) \approx \text{slope at } x_n = \frac{f(x_n) - f(x_{n-1})}{x_n - x_{n-1}} \qquad (3.10)$$

Even though these methods can be implemented in a spreadsheet with relative ease, most equation solver-based software packages feature built-in functions that can return the solutions to such equations more efficiently and accurately, without requiring any programming. Spreadsheet packages such as Excel® also include some built-in functions that are preprogrammed to perform iterative calculations for solving simple equations.

The use of the built-in *Goal Seek* feature of Excel® in solving the problem in Worked Example 3.2 is illustrated in Figure 3.7. In this worksheet, the right-hand side of the equation to be solved is entered in cell C4. Then, the *Goal Seek* option is selected from the *Tools* menu. To start the *Goal Seek* process, cell C4 is specified to be 10 by changing the value of cell C6. The process is instantly executed, and the result is returned as 9.21014.

Alternatively, in equation solver-based software packages, such equations can be solved readily by calling appropriate built-in routines. For example, Figure 3.8 shows how the above problem can be solved in Mathematica®. The variables in the equation are defined first. The built-in routine *Solve* is called where the first argument contains the equation to be solved. The second argument is the variable for which the equation is to be solved. When executed, the solution is returned in line *Out[2]* as 9.21034. Note that the Mathematica® sheet is set up so that one can change the values of the variables and readily solve the equation for *t*. In addition, the same setup can be used to solve for any one of the four variables in the equation, provided the

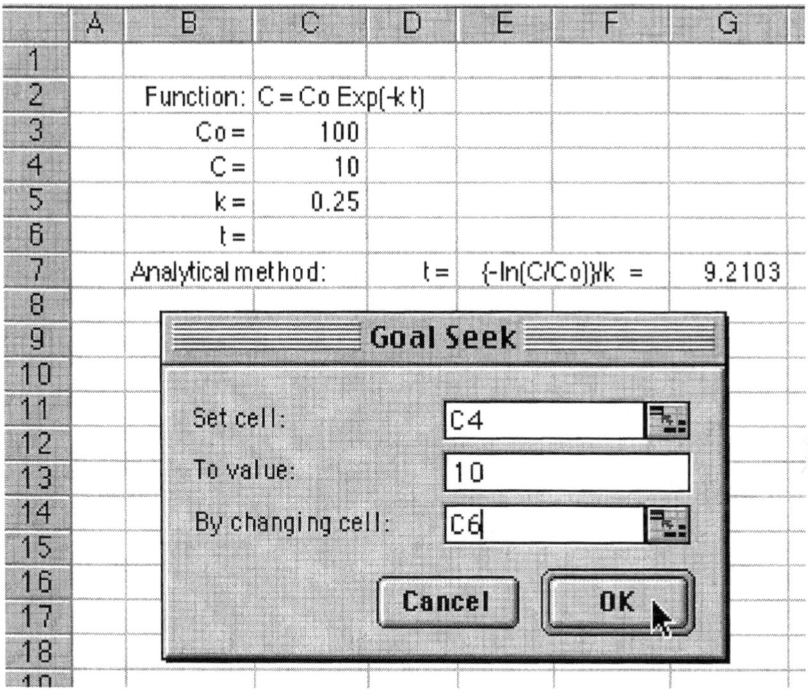

| | A | B | C | D | E | F | G |
|---|---|---|---|---|---|---|---|
| 1 | | | | | | | |
| 2 | | Function: | C = Co Exp(-k t) | | | | |
| 3 | | Co = | 100 | | | | |
| 4 | | C = | 10 | | | | |
| 5 | | k = | 0.25 | | | | |
| 6 | | t = | | | | | |
| 7 | | Analytical method: | | t = | {-ln(C/Co)}/k = | | 9.2103 |
| 8 | | | | | | | |

**Goal Seek**

Set cell: [ C4 ]

To value: [ 10 ]

By changing cell: [ C6 ]

[ Cancel ]  [ OK ]

Figure 3.7 Using *Goal Seek* function in Excel® for solving nonlinear equations.

```
In[1]:= co=100.0; c=10.0; k = 0.25;
        Solve[ c==co*Exp[-k*t], t]
```

```
Out[2]= {{t → 9.21034}}
```

Figure 3.8 Using Mathematica® to solve nonlinear equations.

other three are known. This is achieved by specifying the unknown variable to solve for as the second argument to the call to *Solve*.

### 3.3.1.5 Set of Nonlinear Equations

The general form of a set of nonlinear equations can consist of $n$ functions in terms of $n$ unknown variables, $x_i$:

$$f_1(x_1, x_2, \ldots x_n) = 0; f_2(x_1, x_2, \ldots x_n) = 0; \ldots.$$

and so on up to

$$f_n(x_1, x_2, \ldots x_n) = 0 \tag{3.11}$$

There are no direct methods for solving simultaneous nonlinear equations. The most popular method for solving nonlinear equations is the Newton's Iteration Method, which is based on Taylor's expansion of each of the $n$ equations. For example, the first of the above equations can be expressed as follows:

$$f_1(x_1 + \Delta x_1, \ldots x_n + \Delta x_n) = f_1(x_1, \ldots x_n) + \left(\frac{\partial f_1}{\partial x_1}\right)\Delta x_1$$

$$+ \text{ higher-order terms} \tag{3.12}$$

Neglecting higher-order terms,

$$f_1(x_1 + \Delta x_1, \ldots x_n + \Delta x_n) = f_1(x_1, \ldots x_n) + \left(\frac{\partial f_1}{\partial x_1}\right)\Delta x_1 \tag{3.13}$$

If, from a guessed value of $x1$, the change $\Delta x1$ would make the left-hand side approach 0, then the true root will be $x1 + \Delta x1$. Or, in other words, if

$$\left(\frac{\partial f_1}{\partial x_1}\right)\Delta x_1 = -f_1(x_1, \ldots x_n) \tag{3.14}$$

then $x_1 + \Delta x_1$ will be a root. Thus, extending this argument to the set of equations, this algorithm reduces to the solution of a set of linear equations that can be represented in the following form:

$$
\begin{bmatrix}
\dfrac{\partial f_1}{\partial x_1} & \dfrac{\partial f_1}{\partial x_2} & \cdots & \dfrac{\partial f_1}{\partial x_n} \\[2mm]
\dfrac{\partial f_2}{\partial x_1} & \dfrac{\partial f_2}{\partial x_2} & \cdots & \dfrac{\partial f_2}{\partial x_n} \\[2mm]
\dfrac{\partial f_n}{\partial x_1} & \dfrac{\partial f_n}{\partial x_2} & \cdots & \dfrac{\partial f_n}{\partial x_n}
\end{bmatrix}
\begin{bmatrix}
\Delta x_1 \\[2mm] \Delta x_2 \\ . \\ . \\ . \\ \Delta x_n
\end{bmatrix}
=
\begin{bmatrix}
-f_1 \\[2mm] -f_2 \\ . \\ . \\ . \\ -f_n
\end{bmatrix}
\tag{3.15}
$$

The algorithms discussed in the previous section for a set of linear equations can now be applied to the above set of linear equations to find their roots.

### 3.3.2 ORDINARY DIFFERENTIAL EQUATIONS

A vast majority of environmental systems can be described by ODEs. Only in a limited number of such cases can these equations be solved analytically, whereas all of them can be readily solved using computational methods. Whichever method is used, the solution of ODEs requires that the dependent variable and/or its derivatives be known constraints at prescribed values of the independent variable. When all of these known constraints are at zero value of the independent variable, the problem is called an "initial value problem." In other cases, where the constraints are known at nonzero values of the independent variable, the problem is called a "boundary value problem." The constraints in the two cases are known as "initial conditions" (ICs) and "boundary conditions" (BCs), respectively. It has to be noted that the solution to a differential equation should satisfy the equation and the initial or boundary conditions.

### 3.3.2.1 Analytical Solutions of ODEs

The general form of an ODE may be stated as follows:

$$\frac{dy}{dx} = f(x,y)$$

with an IC of

$$y(x_0) = y_0 \tag{3.16}$$

The solution to this will be a function $y(x)$ that satisfies *both* the differential equation and the IC. Analytical solutions for some of the more common functions $f(x,y)$ and ICs are summarized here. Additional solutions can be found in mathematical handbooks.

*a.* First-order equations: when an ODE can be expressed in the form:

$$M(x)dx + N(y)dy = 0$$

then the equation is *separable,* and the solution can be found by integrating each term in:

$$\int M(x)dx = -\int N(y)dy \tag{3.17}$$

*b.* First-order nonhomogeneous linear equations: when an ODE can be expressed in the form:

$$\frac{dy}{dx} + P(x)y = Q(x)$$

it can be solved by using an integrating factor of the form: $e^{\int P(x)dx}$ to give the solution as

$$y = \frac{\int \{e^{\int P(x)dx}Q(x)\}dx + b}{e^{\int P(x)dx}} \tag{3.18}$$

where $b$ is a constant of integration.

c.  Second-order equation with equidimensional coefficients: when the ODE
    can be expressed in the form:

$$\frac{d^2y}{dx^2} + \frac{a}{x}\frac{dy}{dx} + \frac{b}{x^2}y = 0$$

a trial solution of the form $y = cx^s$, where $s$ is found from:

$$s = -\frac{(a-1) \pm \sqrt{(a-1)^2 - 4b}}{2} \qquad\qquad (3.19)$$

giving the solution as

$$y = b_1x^{s1} + b_2x^{s2} \qquad \text{if } s1 \neq s2 \qquad (3.20a)$$

$$y = b_1x^s + b_2x^s \ln x \qquad \text{if } s1 = s2 = s \qquad (3.20b)$$

d.  Second-order equation with constant coefficients: when the ODE can be
    expressed in the form:

$$\frac{d^2y}{dx^2} + a\frac{dy}{dx} + by = 0$$

a trial solution of the form $y = be^{sx}$, where $s$ is found from:

$$s = -\frac{a \pm \sqrt{a^2 - 4b}}{2} \qquad\qquad (3.21)$$

giving the solution as

$$y = b_1e^{s1x} + b_2e^{s2x} \qquad \text{if } s1 \neq s2 \qquad (3.22a)$$

$$y = b_1e^{sx} + b_2xe^{sx} \qquad \text{if } s1 = s2 = s \qquad (3.22b)$$

### Worked Example 3.3

The buildup of the concentration, $C$, in a lake due to a new step waste load
input can be described by the following equation (to be derived in Chapter 6):

$$\frac{dC}{dt} + 1.23C = 2.0$$

The initial concentration of the waste material in the lake $= 0$. Develop a solution to the above ODE to describe the concentration in the lake as a function
of time.

*Solution*

The solution can be found by first categorizing the ODE and then following standard appropriate mathematical calculi. Alternatively, some computer software packages can find the analytical solution directly. The two approaches are illustrated in this example.

The given equation can be seen as a special case of a first-order non-homogeneous linear equation identified earlier, with $y = C$, $x = t$, $P(x) = 1.23$, and $Q(x) = 2.0$. It can, therefore, be solved by applying Equation (3.18):

$$\text{Integrating factor} = \equiv e^{\int P(t)dt} \equiv e^{\int 1.23 dt} = e^{1.23t}$$

Hence, the solution to ODE is as follows:

$$C = \frac{\int \{e^{1.23t}2\}dt + b}{e^{1.23}t} = \frac{\dfrac{2}{1.23}\{e^{1.23t}\} + b}{e^{1.23t}}$$

$$= \frac{2}{1.23}\left\{1 + \left(\frac{1.23}{2}\right)be^{-1.23t}\right\}$$

From the initial condition given, $C = 0$ at $t = 0$; therefore, $b = -\frac{2}{1.23}$

Thus, the final solution for $C$ as a function of $t$ is as follows:

$$C = \frac{2}{1.23}\{1 - e^{-1.23t}\} = 1.63\{1 - e^{-1.23t}\}$$

See Worked Example 3.4 for a plot of the above result.

As an alternate approach, an equation solver-based software package, Mathematica®, is used to find the solution directly as shown in Figure 3.9. Here, the built-in procedure, *DSolve*, is called in line *In[1]* with the following arguments: the ODE to be solved, the initial condition, the dependent variable, and the independent variable. The solution is returned in line *Out[2]*, with the integration constant automatically evaluated, by Mathematica®. Finally, the *Plot* function is called to plot the solution showing the variation of $C$ with $t$.

It can be seen that the solution returned by Mathematica® in line *Out[2]*, after some manipulation, is identical to the solution found earlier by following the standard mathematical calculi. This example illustrates just one of the benefits of such packages in easing the mathematical tasks involved in modeling, enabling modelers to focus more on formulating and posing the problem at hand in standard mathematical forms rather than on the operandi of solving the formulation.

In[1]:= `ClosedForm=DSolve[{c'[t] + 1.23*c[t]==2.0,`
       `c[0]==0},c,t]`

Out[1]= $\{\{c \to (0.00813008\ e^{-1.23\,\#1}\ (-200.\ +200.\ e^{1.23\,\#1})\ \&)\}\}$

In[2]:= `(c/.ClosedForm[[1,1]])[t]`

Out[2]= $0.00813008\ e^{-1.23\,t}\ (-200.\ +200.\ e^{1.23\,t})$

In[3]:= `Plot[%,{t,0,5}, AxesLabel->{"Time [yrs]" ,`
       `"Conc.in lake [mg/L]"}]`

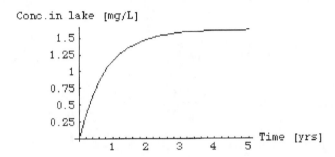

**Figure 3.9** Solution of ODE by Mathematica[®].

### 3.3.2.2 Computational Solutions of ODEs

Two of the more common computational methods used for solving such equations, Euler's Method and the Runge-Kutta Method, are presented here.

(1) Euler's method: this method is based on the application of Taylor's series to estimate the function. Starting from the known initial condition $[x_0, y(x_0)]$, the next point on the function, small step $h$ away from $x$, at $x = x_0 + h$, is found using the slope of the function at $[x_0, y(x_0)]$:

$$y(x_0 + h) = y(x_0) + hy'(x_0) + \frac{h^2}{2}y''(x_0) + \ldots \tag{3.23}$$

By ignoring terms of $h^2$ and higher order, the above can be approximated by the following:

$$y(x_0 + h) \cong y(x_0) + hy'(x_0) \tag{3.24}$$

This procedure may be continued for $n$ points along the function for $n = 1,2,3, \ldots$ to develop the function $y(x)$ over the desired range:

$$y_{n+1} = y_n + hf'(x_n, y_n) \tag{3.25}$$

**Worked Example 3.4**

Solve the ODE from Worked Example 3.3 using Euler's method, and compare the result with the analytical solution found in Worked Example 3.3.

*Solution*

Euler's numerical method is illustrated in Figure 3.10 using a time step, $h = 0.05$ year. The numerical method, in this case, is rather close to the analytical solution. The error will depend on the time step and the type of equation being solved.

(2) Runge-Kutta method: the error in Euler's method stems from the assumption that the slope at the beginning of the calculation step is the same

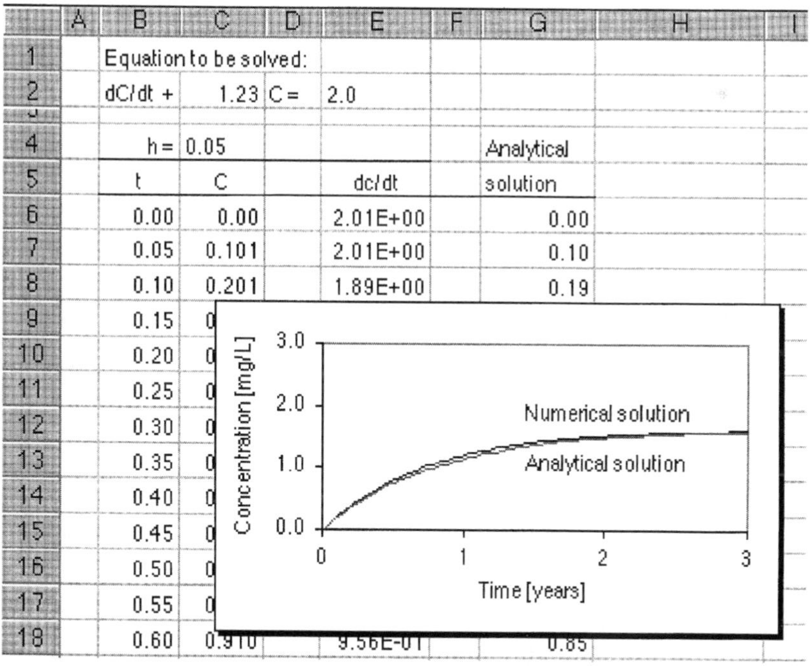

| | A | B | C | D | E | F | G | H | I |
|---|---|---|---|---|---|---|---|---|---|
| 1 | | Equation to be solved: | | | | | | | |
| 2 | | dC/dt + | 1.23 | C = | 2.0 | | | | |
| 4 | | h = | 0.05 | | | | Analytical | | |
| 5 | | t | C | | dc/dt | | solution | | |
| 6 | | 0.00 | 0.00 | | 2.01E+00 | | 0.00 | | |
| 7 | | 0.05 | 0.101 | | 2.01E+00 | | 0.10 | | |
| 8 | | 0.10 | 0.201 | | 1.89E+00 | | 0.19 | | |
| 9 | | 0.15 | 0 | | | | | | |
| 10 | | 0.20 | 0 | | | | | | |
| 11 | | 0.25 | 0 | | | | | | |
| 12 | | 0.30 | 0 | | | | | | |
| 13 | | 0.35 | 0 | | | | | | |
| 14 | | 0.40 | 0 | | | | | | |
| 15 | | 0.45 | 0 | | | | | | |
| 16 | | 0.50 | 0 | | | | | | |
| 17 | | 0.55 | 0 | | | | | | |
| 18 | | 0.60 | 0.910 | | 9.56E-01 | | 0.85 | | |

**Figure 3.10** Euler's method vs. analytical solution.

over the entire step, which may be considered a first-order method. Under the Runge-Kutta method, two schemes have been proposed to improve the accuracy—a second-order method and a fourth-order method. The latter is the most commonly used in environmental modeling practice.

In the Runge-Kutta method, higher-order terms in the Taylor series are retained. The formulas for the fourth-order method are as follows:

$$y_{n+1} = y_n + \frac{(K_0 + 2K_1 + 2K_2 + K_3)}{6} \tag{3.26}$$

where

$$K_0 = hf(x_n, y_n)$$

$$K_1 = hf(x_n + 0.5h, y_n + 0.5K_0)$$

$$K_2 = hf(x_n + 0.5h, y_n + 0.5K_1)$$

$$K_3 = hf(x_n + h, y_n + K_2)$$

Even though the fourth-order Runge-Kutta method is more computation intensive, because of its increased accuracy, higher step sizes, $h$, can be used. The above algorithm can be implemented in spreadsheet packages, such as Excel®, but can be quite tedious; however, most computer software packages have the Runge-Kutta routines built in, and they can be evoked readily with appropriate arguments. An example of the Excel® implementation of the Runge-Kutta method is illustrated in Worked Example 3.5.

### Worked Example 3.5

Solve the ODE from the Worked Example 3.3 using the Runge-Kutta method.

### Solution

The Runge-Kutta method implemented in the Excel® spreadsheet package is shown in Figure 3.11. The comparison between the solution by the Runge-Kutta method and the analytical solution is also included, showing close agreement between the two.

Equation to solve:  dC/dt =  2.0 - 1.23 C

Initial conditions:

    t0= 0        Co = 0

Analytical solution:

    C= 1.63*(1-exp(-1.23 * t))

Numerical Solution by Runge-Kutta Method:

Final value of X:         Number of steps:      Step size:

  xf = 3            N = 60           h = 0.05

| t | t+h/2 | K1 | C+hK1/2 | K2 | C+hK2/2 | K3 | C+hK3 | K4 | Num. solution | Anal. solution |
|---|---|---|---|---|---|---|---|---|---|---|
| 0.00 | | | | | | | | | 0.000 | 0.000 |
| 0.05 | 0.0 | 2.000 | 0.050 | 2.000 | 0.050 | 2.000 | 0.100 | 2.000 | 0.100 | 0.097 |
| 0.10 | 0.1 | 1.877 | 0.147 | 1.877 | 0.147 | 1.877 | 0.194 | 1.877 | 0.194 | 0.188 |
| 0.15 | 0.1 | 1.762 | 0.238 | 1.762 | 0.238 | 1.762 | 0.282 | 1.762 | 0.282 | 0.274 |
| 0.20 | 0.2 | 1.653 | | | | | | | | 0.355 |
| 0.25 | 0.2 | 1.552 | | | | | | | | 0.430 |
| 0.30 | 0.3 | 1.456 | | | | | | | | 0.502 |
| 0.35 | 0.3 | 1.367 | | | | | | | | 0.569 |
| 0.40 | 0.4 | 1.283 | | | | | | | | 0.632 |
| 0.45 | 0.4 | 1.204 | | | | | | | | 0.691 |
| 0.50 | 0.5 | 1.130 | | | | | | | | 0.747 |
| 0.55 | 0.5 | 1.060 | | | | | | | | 0.799 |
| 0.60 | 0.6 | 0.995 | | | | | | | | 0.849 |
| 0.65 | 0.6 | 0.934 | | | | | | | | 0.895 |
| 0.70 | 0.7 | 0.876 | | | | | | | | 0.939 |
| 0.75 | 0.7 | 0.822 | 0.978 | 0.822 | 0.978 | 0.822 | 0.998 | 0.822 | 0.998 | 0.980 |

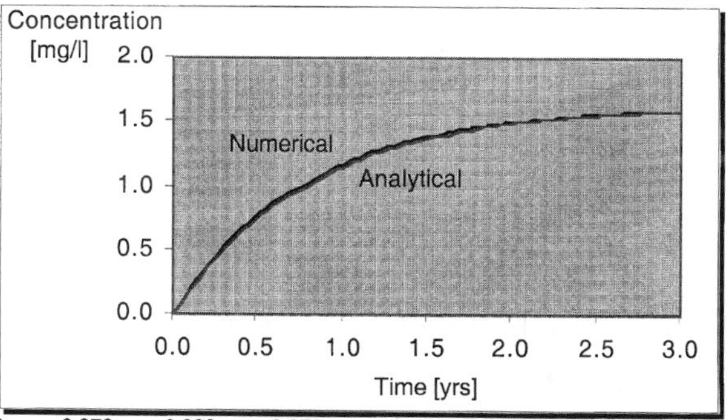

Concentration [mg/l] — Numerical, Analytical; Time [yrs] (0.0 to 3.0)

**Figure 3.11** Runge-Kutta method in Excel®.

## 3.3.3 SYSTEMS OF ORDINARY DIFFERENTIAL EQUATIONS

Systems of coupled ODEs are rather common in many environmental systems and several examples will be considered in detail in the following chapters. The Runge-Kutta method can be extended to systems of ODEs as well as to higher-order differential equations. Higher-order equations have to be reduced to first-order equations by introducing new variables before applying the Runge-Kutta method. For example, consider the second-order differential equation:

$$\frac{d^2y}{dx^2} = g\left(x, y, \frac{dy}{dx}\right) \qquad (3.27)$$

The order of the above can be reduced from 2 to 1 by introducing

$$z = \frac{dy}{dx} \quad \text{whereby} \quad \frac{dz}{dx} = \frac{d^2y}{dx^2} \qquad (3.28)$$

Hence, the original problem is now equivalent to one with two coupled ODEs:

$$\frac{dz}{dx} = g(x,y,z) \quad \text{and} \quad \frac{dy}{dx} = f(x,y,z) = z \tag{3.29}$$

The Runge-Kutta method can now be applied to each of the above two ODEs:

$$y_{n+1} = y_n + \frac{(K_0 + 2K_1 + 2K_2 + K_3)}{6} \tag{3.30}$$

$$z_{n+1} = z_n + \frac{(L_0 + 2L_1 + 2L_2 + L_3)}{6} \tag{3.31}$$

where

$$K_0 = hf(x_n, y_n, z_n)$$
$$L_0 = hg(x_n, y_n, z_n)$$
$$K_1 = hf(x_n + 0.5h, y_n + 0.5K_0, z_n + 0.5L_0)$$
$$L_1 = hg(x_n + 0.5h, y_n + 0.5K_0, z_n + 0.5L_0)$$
$$K_2 = hf(x_n + 0.5h, y_n + 0.5K_1, z_n + 0.5L_1)$$
$$L_2 = hg(x_n + 0.5h, y_n + 0.5K_1, z_n + 0.5L_1)$$
$$K_3 = hf(x_n + h, y_n + K_2, z_n + L_2)$$
$$L_3 = hg(x_n + h, y_n + K_2, z_n + L_2)$$

The above algorithms are available in many software packages as built-in routines for easy application. Several examples of such applications to a wide range of problems will be presented in the following chapters.

### 3.3.4 PARTIAL DIFFERENTIAL EQUATIONS

For a large variety of environmental systems, the dependent variable is expressed in more than one independent variable. For example, the concentration of a pollutant in a river may be a function of time and river miles, whereas that in an aquifer may be a function of the three spatial dimensions as well as of time. Such systems are modeled using partial differential equations.

Partial differential equations may be classified in terms of their mathematical form or the type of problem to which they apply. The general form of a second-order PDE in two independent variables is as follows:

$$A(x,y)\frac{\partial^2 f}{\partial x^2} + B(x,y)\frac{\partial^2 f}{\partial x \partial y} + C(x,y)\frac{\partial^2 f}{\partial y^2} + E\left(x, y, \frac{\partial f}{\partial x}, \frac{\partial f}{\partial y}\right) \tag{3.32}$$

If $B^2 - 4AC < 0$, the equation is classified as elliptic; if $B^2 - 4AC = 0$, parabolic; and if $B^2 - 4AC < 0$, hyperbolic. To solve such PDEs, initial conditions, boundary conditions, or their combinations would be required. However, most PDEs cannot be solved in exact form analytically; analytical solutions, when available, are problem specific and cannot be generalized. Therefore, numerical approaches are the methods of choice for their solution, of which the finite difference method and the finite element method are the most common.

### 3.3.4.1 Finite Difference Method

The essence of this method is as follows. The problem domain is first divided into a grid of $n$ node points. The PDE is approximated by difference equations relating the functional value at neighboring points in the grid. Then, the resulting set of $n$ equations in $n$ unknowns is solved to obtain the approximate solution values at the node points. The approximation of PDEs by difference is based on Taylor series expansion and can be formulated by considering the two-dimensional grid system shown in Figure 3.12.

Using the notation of subscript "$j$" for the $y$-variable and subscript "$i$" for the $x$-variable, the difference equations take the following form:

$$\frac{\partial f}{\partial x} = \frac{f_{i+1,j} - f_{i-1,j}}{2h}; \quad \frac{\partial f}{\partial y} = \frac{f_{i,j+1} - f_{i,j-1}}{2h}; \tag{3.33}$$

$$\frac{\partial^2 f}{\partial x^2} = \frac{f_{i+1,j} - 2f_{i,j} + f_{i-1,j}}{h^2}; \quad \frac{\partial^2 f}{\partial y^2} = \frac{f_{i,j+1} - 2f_{i,j} + f_{i,j-1}}{h^2}; \tag{3.34}$$

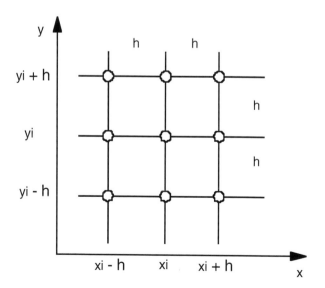

**Figure 3.12** Two-dimensional grid system.

$$\frac{\partial^2 f}{\partial x \partial y} = \frac{f_{i+1,j+1} - f_{i-1,j+1} - f_{i+1,j-1} + f_{i-1,j-1}}{4h^2} \qquad (3.35)$$

***Worked Example 3.6***

The advective-diffusive transport of a pollutant in a river, undergoing a first-order decay reaction, can be modeled by the following equation (as derived in Chapter 2):

$$\frac{\partial C}{\partial t} = E\frac{\partial^2 C}{\partial t^2} - U\frac{\partial C}{\partial x} - kC$$

where $C$ is the concentration of the pollutant, $E$ is the dispersion coefficient, $U$ is the velocity, $k$ is the reaction rate constant, $t$ is the time, and $x$ is the distance along the river. Develop the finite difference formulation to solve the above PDE numerically.

*Solution*

The grid notation indicated in Figure 3.12 can be adapted as shown in Figure 3.13 for this problem. Each of the terms in the PDE can now be expressed in the finite difference form as follows:

$$\frac{\partial C}{\partial t} = \frac{C_{i,n+1} - C_{i,n}}{h}$$

$$E\frac{\partial^2 C}{\partial x^2} = E\left(\frac{C_{i+1,n} - 2C_{i,n} + C_{i-1,n}}{h^2}\right)$$

$$U\frac{\partial C}{\partial x} = U\left(\frac{C_{i,n} - C_{i-1,n}}{h}\right)$$

$$kC = k\left(\frac{C_{i,n} + C_{i,n+1}}{2}\right)$$

Hence, combining all of the above, the finite difference form of the PDE is

$$\frac{C_{i,n+1} - C_{i,n}}{h} = E\left(\frac{C_{i+1,n} - 2C_{i,n} + C_{i-1,n}}{h^2}\right) - U\left(\frac{C_{i,n} - C_{i-i,n}}{h}\right)$$

$$- k\left(\frac{C_{i,n} + C_{i,n+1}}{2}\right)$$

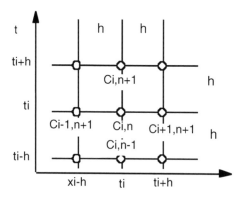

**Figure 3.13**

which can be rearranged to solve for $C_{i,n+1}$ in terms of the known values of $C$ at the $n$th time point, $C_{i,n}$. In essence, these equations now become a set of algebraic equations to be solved simultaneously.

Even though the above equations can be set up in a spreadsheet package, the implementation can be laborious. Several mathematical packages offer built-in routines to execute this algorithm, without requiring any programming on the part of the modeler.

## 3.4 CLOSURE

Mathematical calculi underlying various types of mathematical formulations were presented in this chapter. These formulations are integral components of mathematical models, and a clear understanding of these formulations and the calculi is essential to complete the modeling process. The availability of modern software packages can be of significant benefit to modelers, because the mechanics of the mathematical calculi are readily available, preprogrammed. The intent of this book is to demonstrate, through specific examples, that many of these calculi could be implemented in a computer environment using software packages that require minimal programming skills. These packages can, thus, help subject matter experts focus on formulating and posing the problem rather than on implementing the problem and the solution procedure in a computer environment to achieve the modeling goal.

It should be noted that some software packages incorporate the calculi and algorithms as built-in ready-to-use "libraries"; in some other packages, the users may have to chose from several built-in algorithms; in yet others, users may have to set them up with some software-specific scripts, syntax, or code. This is the reason for including this mathematics primer in this book, so that readers can benefit from them in selecting the optimal software package for modeling, in diagnosis and troubleshooting, and in verifying and interpreting the results obtained.

## EXERCISE PROBLEMS

3.1. The relationship between pressure, $P$, volume, $V$, and temperature, $T$, for many real gases can be described by the van der Waals equation:

$$\left(P + \frac{a}{V^2}\right)(V - b) = RT$$

Identify the problems with solving the above equation using elementary algebraic methods. Prepare an Excel® worksheet to solve this equation for $V$ for known values of $P$, $T$, $a$, and $b$.

3.2. The rate expressions for the reversible reaction $A \leftrightarrows B$ result in the following coupled differential equations:

$$\frac{d[A]}{dt} = -k_1[A] + k_2[B] \quad \text{and} \quad \frac{d[B]}{dt} = k_1[A] - k_2[B]$$

Recognizing that the total concentration of A and $B = C_T$, the rate equation for A can be found as follows:

$$\frac{d[A]}{dt} = -k_1[A] + k_2[C_T - A]$$

Show that the solution to the above ODE is

$$A = A_0 \exp[-k_1 + k_2)t] + \frac{k_2 C_T}{k_1 + k_2} \{1 - \exp[-(k_1 + k_2)t]\}$$

where $A_0$ is the initial concentration of A.

Assuming $k_1 = 0.2$, $k_2 = 0.1$, $C_T = 0.01$, and $A_0 = 0.009$, find the time required for A to decrease by 50%.

3.3. Consecutive reactions $A \rightarrow B \rightarrow C$ such as can be found in environmental systems (e.g., $NH_3 \rightarrow NO_2^- \rightarrow NO_3^-$). Starting from the individual rate equations:

$$\frac{d[A]}{dt} = -k_1[A]; \quad \frac{d[B]}{dt} = k_1[A] + k_2[B]$$

and

$$\frac{d[C]}{dt} = k_3[B]$$

deduce that the rate equation for B can be written as follows:

$$\frac{d[B]}{dt} = k_1 A_0 \exp(-k_1 t) + k_2[B]$$

Hence, show that

$$B = B_0 \exp(-k_2 t) + \frac{A_0 k_1}{k_2 - k_1}[\exp(-k_1 t) - \exp(-k_2 t)]$$

where $A_0$ and $B_0$ are the initial concentrations of A and B.

Assuming $k_1 = 0.15$, $k_2 = 0.2$, $A_0 = 0.05$, and $B_0 = 0.01$, find the time required for B to increase by 50%.

3.4. The rate of transfer of oxygen in natural waters by reaeration alone can be modeled by the following equation:

$$\frac{dC}{dt} = K_L \frac{A}{V}(C_{sat} - C)$$

Solve the above ODE analytically, as it is and also after making a substitution, $D = (C_{sat} - C)$, and confirm that both approaches yield identical results for C as a function of time.

Develop an Excel® worksheet to solve the equation numerically, and compare the results with the analytical results by plotting a graph of C vs. t.

Assume $(K_L A/V) = 2.5$ day$^{-1}$ and $C_{sat} = 7.8$ mg/L.

3.5. The in-stream concentration of a pollutant caused by a linearly distributed uniform source of strength $S_D$ can be described by the following equation:

$$E\frac{d^2C}{dx^2} - U\frac{dC}{dx} - KC = S_D$$

that gives the following solution:

For the reach $x \leq 0$: $C = \left\{ \frac{S_D}{K}\left(\frac{\alpha - 1}{2\alpha}\right)(1 - \exp(-j_1 a)) \right\} \exp(j_1 x)$

For the reach $0 \leq x \leq a$: $C = \frac{S_D}{K}\left[ 1 - \left(\frac{\alpha - 1}{2\alpha}\right) \exp(j_1(x - a)) \right.$

$$\left. - \left(\frac{\alpha + 1}{2\alpha}\right) \exp(j_2(x)) \right]$$

For the reach $x \geq a$: $C = \left\{ \dfrac{S_D}{K} \left( \dfrac{\alpha + 1}{2\alpha} \right)[1 - \exp(-j_2 a)] \right\} \exp(j_2(x - a))$

where the distributed load occurs from $x = 0$ to $x = a$. Determine the optimum value of $a$ over which the waste load can be distributed without violating the following water quality standards for the pollutant: at $x \leq -6$ miles, $C$ should be less than 1.5 mg/L and at $x \geq 40$ miles, $C$ should be less than 2.0 mg/L.

The maximum value of $C$ in the stream should be less than 20 mg/L. (Hint: use the *Goal Seek* or *Solver* feature in Excel®).

Assume the following values: $E = 15$ miles$^2$/day, $U = 4$ miles/day, $K = 0.3$ day$^{-1}$, $S_D = 25$ mg/L-day, $\alpha = 1.44$, $j_1 = 0.33$ mile$^{-1}$, and $j_2 = 0.06$ mile$^{-1}$.

## CHAPTER 4

# Fundamentals of Environmental Processes

## CHAPTER PREVIEW

*The objective of this chapter is to review the fundamental principles required to formulate material balance equations, which are the building blocks in mathematical modeling of environmental systems. Toward this end, fundamental concepts and principles of environmental processes commonly encountered in both engineered and natural systems are reviewed here. Topics reviewed include phase contents, phase equilibrium, partitioning, transport processes, and reactive and nonreactive processes. Finally, integration of these concepts in formulating the material balance equation is outlined. Again, rather than rigorous and complete thermodynamic and mechanistic analyses, only the extracts are included to serve the modeling goal.*

## 4.1 INTRODUCTION

T HE ultimate objective of this book is to develop models to describe the changes of concentrations of contaminants in engineered and natural systems. Changes within a system can result due to transport into and/or out of the system and/or processes acting on the contaminants within the system. Contaminants can be transported through a system by microscopic and macroscopic mechanisms such as diffusion, dispersion, and advection. At the same time, they may or may not undergo a variety of physical, chemical, and biological processes within the system. Some of these processes result in changes in the molecular nature of the contaminants, while others result in mere change of phase or separation. The former type of processes can be categorized as *reactive processes* and the latter as *nonreactive processes*. Those contaminants that do not undergo any significant processes are called *conservative* substances, and those that do are called *reactive* substances. While most contaminants are reactive to a good extent, a few substances behave as conservative substances. Nonreactive substances such as chlorides can be used as tracers to study certain system characteristics.

A clear mechanistic understanding of the processes that impact the fate and transport of contaminants is a prerequisite in formulating the mathematical model. At the microscopic level, such processes essentially involve the same reaction and mass transfer considerations irrespective of the system (engineered or natural) or the media (soil, water, or air) or the phase (solid, liquid, or gas) in which they occur. The system will bring in additional specific considerations at the macroscopic level. Basic definitions and fundamental concepts relating to the microscopic level processes common to both engineered and natural systems are reviewed in this chapter. Express details pertinent to engineered and natural systems are reviewed in Chapters 5 and 6, respectively. The specific objective here is to compile general expressions or *submodels* for quantifying the rate of mass "transferred" or "removed" by the various processes that would cause changes of concentrations in the system.

## 4.2 MATERIAL CONTENT

Material content is a measure of the material contained in a bulk medium, quantified by the ratio of the amount of material present to the amount of the medium. The amounts can be quantified in terms of mass, moles, or volume. Thus, the ratio can be expressed in several alternate forms such as mass or moles of material per volume of medium resulting in mass or molar concentration; moles of material per mole of medium, resulting in mole fraction; volume of material per volume of medium, resulting in volume fraction; and so on. The use of different forms of measures in the ratio to quantify material content may become confusing in the case of mixtures of materials and media. The following notation and examples can help in formalizing these different forms: subscripts for components are $i = 1,2,3, \ldots N;$ and subscripts for phases are $g$ = gas, $a$ = air, $l$ = liquid, $w$ = water, $s$ = solids and soil.

### 4.2.1 MATERIAL CONTENT IN LIQUID PHASES

Material content in liquid phases is often quantified as mass concentration, molar concentration, or mole fraction.

$$\text{Mass concentration of component } i \text{ in water} = \rho_{i,w} = \frac{\text{Mass of material, } i}{\text{Volume of water}} \quad (4.1)$$

Molar concentration of component $i$ in water $= C_{i,w} = \dfrac{\text{Moles of material, } i}{\text{Volume of water}}$ (4.2)

Because moles of material = mass ÷ molecular weight, $MW$, mass concentrations, $\rho_{i,w}$, and molar concentrations, $C_{i,w}$, are related by the following:

$$C_{i,w} = \frac{\rho_{i,w}}{MW_i}$$ (4.3)

Mole fraction, $X$, of a single chemical in water can be expressed as follows:

$$X = \frac{\text{Moles of chemical}}{\text{Moles of chemical + Moles of water}}$$

For dilute solutions, the moles of chemical in the denominator of the above can be ignored in comparison to the moles of water, $n_w$, and $X$ can be approximated by:

$$X = \frac{\text{Moles of chemical}}{\text{Moles of water}}$$ (4.4)

An aqueous solution of a chemical can be considered dilute if $X$ is less than 0.02. Similar expressions can be formulated on mass basis to yield mass fractions. Mass fractions can also be expressed as a percentage or as other ratios such as parts per million (ppm) or parts per billion (ppb).

In the case of solutions of mixtures of materials, it is convenient to use mass or mole *fractions,* because the sum of the individual fractions should equal 1. This constraint can reduce the number of variables when modeling mixtures of chemicals. Mole fraction, $X_i$, of component $i$ in an N-component mixture is defined as follows:

$$X_i = \frac{\text{Moles of } i}{\left( \sum_1^N n_i \right) + n_w}$$ (4.5)

and, the sum of all the mole fractions $= \left( \sum_1^N X_i \right) + X_w = 1$ (4.6)

As in the case of single chemical systems, for dilute solutions of multiple chemicals, mole fraction $X_i$ of component $i$ in an N-component mixture can be approximated by the following:

$$X_i = \frac{\text{Moles of } i}{n_w}$$ (4.7)

This ratio of quantities is independent of the system and the mass of the sample. Such a property that is independent of the mass of the sample is known as an intensive property. Other examples of intensive properties include pressure, density, etc. Those that depend on mass, volume and potential energy, for example, are called extensive properties.

## 4.2.2 MATERIAL CONTENT IN SOLID PHASES

The material content in solid phases is often quantified by a ratio of masses and is expressed as ppm or ppb. For example, a quantity of a chemical adsorbed onto a solid adsorbent is expressed as mg of adsorbate per kg of adsorbent.

## 4.2.3 MATERIAL CONTENT IN GAS PHASES

The material content in gas phases is often quantified by a ratio of moles or volumes and is expressed as ppm or ppb. It is important to specify the temperature and pressure in this case, because (unlike liquids and solids) gas phase densities are strong functions of temperature and pressure. It is preferable to report gas phase concentrations at standard temperature and pressure (STP) conditions of $0°C$ and 760 mm Hg.

***Worked Example 4.1***

A certain chemical has a molecular weight of 90. Derive the conversion factors to quantify the following: (1) 1 ppm (volume/volume) of the chemical in air in molar and mass concentration form, (2) 1 ppm (mass ratio) of the chemical in water in mass and molar concentration form, and (3) 1 ppm (mass ratio) of the chemical in soil in mass ratio form.

*Solution*

(1) In the air phase, the volume ratio of 1 ppm can be converted to the mole or mass concentration form using the assumption of Ideal Gas, with a molar volume of 22.4 L/gmole at STP conditions (273 K and 1.0 atm.).

$$1\ \text{ppm}_v = \frac{1\ m^3\ \text{of chemical}}{1,000,000\ m^3\ \text{of air}}$$

$$1\ \text{ppm}_v \equiv \frac{1\ m^3\ \text{of chemical}}{1,000,000\ m^3\ \text{of air}} \left(\frac{\text{moles}}{22.4\ L}\right)\left(\frac{1000\ L}{m^3}\right) \equiv 4.46 \times 10^{-5}\ \frac{\text{moles}}{m^3}$$

$$\equiv 4.46 \times 10^{-5}\ \frac{\text{moles}}{m^3} \left(\frac{90\ g}{\text{gmole}}\right) \equiv 0.004\ \frac{g}{m^3} \equiv 4\ \frac{mg}{m^3} \equiv 4\ \frac{\mu g}{L}$$

The general relationship is 1 ppm = $(MW/22.4)$ mg/m$^3$.

(2) In the water phase, the mass ratio of 1 ppm can be converted to mole or mass concentration form using the density of water, which is 1 g/cc at 4°C and 1 atm:

$$1 \text{ ppm} = \frac{1 \text{ g of chemical}}{1,000,000 \text{ g of water}}$$

$$1 \text{ ppm} \equiv \frac{1 \text{ g of chemical}}{1,000,000 \text{ g of water}} \left(1 \frac{g}{cm^3}\right)\left(\frac{100^3 cm^3}{m^3}\right) \equiv 1 \frac{g}{m^3} \equiv 1 \frac{mg}{L}$$

$$\equiv 1\frac{g}{m^3}\left(\frac{mole}{90 \text{ g}}\right) \equiv 0.011 \frac{moles}{m^3}$$

(3) In the soil phase, the conversion is direct:

$$1 \text{ ppm} = \frac{1 \text{ g of chemical}}{1,000,000 \text{ g of soil}}$$

$$1 \text{ ppm} = \frac{1 \text{ g of chemical}}{1,000,000 \text{ g of soil}} \left(\frac{1000 \text{ g}}{kg}\right)\left(\frac{1000 \text{ mg}}{g}\right) = 1 \frac{mg}{kg}$$

## Worked Example 4.2

Analysis of a water sample from a lake gave the following results: volume of sample = 2 L, concentration of suspended solids in the sample = 15 mg/L, concentration of a dissolved chemical = 0.01 moles/L, and concentration of the chemical adsorbed onto the suspended solids = 500 μg/g solids. If the molecular weight of the chemical is 125, determine the total mass of the chemical in the sample.

## Solution

Total mass of chemical can be found by summing the dissolved mass and the adsorbed mass. Dissolved mass can be found from the given molar concentration, molecular weight, and sample volume. The adsorbed mass can be found from the amount of solids in the sample and the adsorbed concentration. The amount of solids can be found from the concentration of solids in the sample.

Dissolved concentration = molar concentration * *MW*

$$= 0.001 \frac{moles}{L}\left(\frac{125 \text{ g}}{gmole}\right) = 0.125 \frac{g}{L}$$

Dissolved mass in sample = dissolved concentration × volume

$$= \left(0.125 \, \frac{g}{L}\right) \times (2 \text{ L}) = 0.25 \text{ g}$$

Mass of solids in sample = concentration of solids × volume

$$= \left(25 \, \frac{mg}{L}\right) \times (2 \text{ L}) = 50 \text{ mg} = 0.05 \text{ g}$$

Adsorbed mass in sample = adsorbed concentration × mass of solids

$$= \left(500 \, \frac{\mu g}{g}\right) \times (0.05 \text{ g})\left(\frac{g}{10^6 \, \mu g}\right) = 0.00025 \text{ g}$$

Hence, total mass of chemical in the sample = 0.25 g + 0.00025 g = 0.25025 g.

## 4.3 PHASE EQUILIBRIUM

The concept of phase equilibrium is an important one in environmental modeling that can be best illustrated through an experiment. Consider a sealed container consisting of an air-water binary system. Suppose a mass, $m$, of a chemical is injected into this closed system, and the system is allowed to reach equilibrium. Under that condition, some of the chemical would have partitioned into the aqueous phase and the balance into the gas phase, assuming negligible adsorption onto the walls of the container. The chemical content in the aqueous and gas phases are now measured (as mole fractions, $X$ and $Y$). The experiment is then repeated several times by injecting different amounts of the chemical each time and measuring the final phase contents in each case ($X$'s and $Y$'s). A rectilinear plot of $Y$ vs. $X$, called the *equilibrium diagram*, is then generated from the data, as illustrated in Figure 4.1.

For most chemicals, when the aqueous phase content is dilute, a linear relationship could be observed between the phase contents, $Y$ and $X$. (A commonly accepted criterion for dilute solution is aqueous phase mole fraction, $X < 2\%$.) This phenomenon is referred to as *linear partitioning*. The slope of the straight line in the equilibrium diagram is a temperature-dependent thermodynamic property of the chemical and is termed the *partition coefficient*. Such linearity has been observed for most chemicals in many two-phase environmental systems. Thus, $X$ and $Y$ are related to one another under dilute conditions by

$$Y = K_{a-w}X \tag{4.8}$$

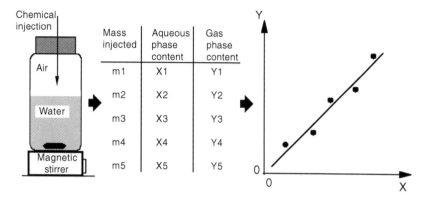

**Figure 4.1** Linear partitioning in air-water binary system.

where, $K_{a-w}$ is the nondimensional air-water partition coefficient (−). Similar partitioning phenomena can be observed between other phases as well. Some of the more common two-phase environmental systems and the appropriate partition coefficients for those systems are summarized in Appendix 4.1. It is imperative that these definitions be used consistently to avoid confusion about units and inverse ratios, i.e., $K_{1-2}$ vs. $K_{2-1}$.

Experimentally measured data for many of these partition coefficients can be found in handbooks and the literature. Alternatively, structure activity relationship (SAR) or property activity relationship (PAR) methods have also been proposed to estimate them from molecular structures or other physico-chemical properties. A comprehensive compilation of such estimation methods can be found in Lyman et al. (1982).

## 4.3.1 STEADY STATE AND EQUILIBRIUM

The concept of steady state has been referred to previously, implying no changes with passage of time. The equilibrium conditions discussed above also imply no change of state with passage of time. The following illustration adapted from Mackay (1991) provides a clear understanding of the similarities and differences between the two concepts.

Consider the oxygen concentrations in the water and air, first, in a closed air-water binary system as shown in Figure 4.2(a). After a sufficiently long time, the system will reach equilibrium conditions with an oxygen content of $8.6 \times 10^{-3}$ mole/L and $2.9 \times 10^{-4}$ mole/L in the gas and aqueous phase, respectively. The system will remain under these conditions, seen as steady state. Consider now the flow system in Figure 4.2(b). The flow rates remain constant with time, keeping the oxygen contents the same as before. The system not only is at steady state, but also is at equilibrium, because the ratio of

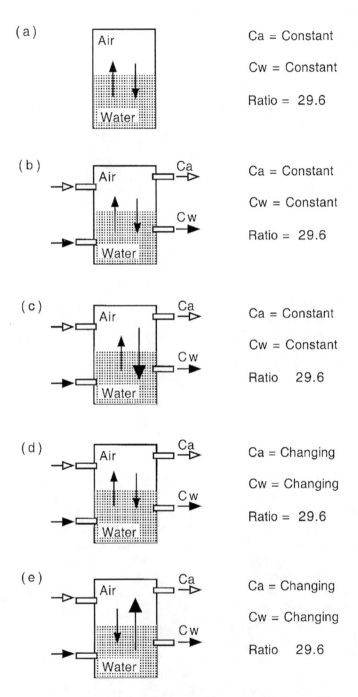

**Figure 4.2** Illustration of steady state conditions vs. equilibrium conditions.

the phase contents is still equal to the $K_{a-w}$ value. Now consider the situation in Figure 4.2(c), where the flow rates are still steady, but the phase contents are not being maintained at the "equilibrium values," and their ratio is not equal to the $K_{a-w}$ value. Here, the system is at steady state but not at equilibrium. In Figure 4.2(d), the flow rates and phase contents are fluctuating with time; however, their ratio remains the same at $K_{a-w}$. Here, the system is not at steady state, but it is at equilibrium. Finally, in Figure 4.2(e), the flow rates and the phase contents and their raito are changing. This system is not at steady state or equilibrium.

## 4.3.2 LAWS OF EQUILIBRIUM

Several fundamental laws from physical chemistry and thermodynamics can be applied to environmental systems under certain conditions. These laws serve as important links between the state of the system, chemical properties, and their behavior. As pointed out earlier, fundamental laws of science form the building blocks of mathematical models. As such, some of the important laws essential for modeling the fate and transport of chemicals in natural and engineered environmental systems are reviewed in the next section.

### 4.3.2.1 Ideal Gas Law

The Ideal Gas Law states that

$$pV = nRT \qquad (4.9)$$

where $p$ is the pressure, $V$ is the volume, $n$ is the number of moles, $R$ is the Ideal Gas Constant, and $T$ is the absolute temperature. Most gases in environmental systems can be assumed to obey this law. It is important to use the appropriate value for $R$ depending on the units used for the other parameters as summarized in Table 4.1.

Table 4.1 Units Used in the Ideal Gas Law

| Pressure, $p$ | Volume, $V$ | Temperature, $T$ | No. of Moles, $n$ | Ideal Gas Constant, $R$ |
|---|---|---|---|---|
| atm. | L | K | gmole | 0.08206 atm.-L/gmole-K |
| mm Hg | L | K | gmole | 62.36 mm Hg-L/gmole-K |
| atm. | $ft^3$ | K | lbmole | 1.314 atm.-$ft^3$/lbmole-K |
| psi | $ft^3$ | R | lbmole | 10.73 psi.-$ft^3$/lbmole-R |
| in Hg | $ft^3$ | R | lbmole | 21.85 in Hg-$ft^3$/lbmole-R |

### 4.3.2.2 Dalton's Law

Dalton's Law states that for an ideal mixture of gases of total volume, $V$, the total pressure, $p$, is the sum of the *partial pressures*, $p_i$, exerted by each component in the mixture. Partial pressure is the pressure that would be exerted by the component if it occupied the same total volume, $V$, as that of the mixture. The following relationships can be developed for an N-component mixture of ideal gases:

$$p = \sum_{j=1}^{N} p_j = p_A + p_B + p_C \ldots \tag{4.10}$$

and,

$$p_j = \frac{n_j RT}{V} \tag{4.11}$$

where $n_j$ is the number of moles of component $j$ in the mixture. A useful corollary can be deduced by combining the above two equations:

$$p = p_A + p_B + p_C \ldots = \frac{(n_A + n_B + n_C \ldots)RT}{V} \tag{4.12}$$

Considering component $A$, as an example, its mole fraction in the mixture, $Y_A$, can now be related to its partial pressure as follows:

$$Y_A = \frac{n_A}{n_A + n_B + n_C \ldots} = \frac{\dfrac{p_A V}{RT}}{\dfrac{pV}{RT}}$$

or

$$Y = \frac{p_A}{p} \tag{4.13}$$

### 4.3.2.3 Raoult's Law

Raoult's Law states that the partial pressure, $p_A$, of a chemical $A$ in the gas phase just above a liquid phase containing the dissolved form of the chemical $A$ along with other chemicals, is given by

$$p_A = vp_A X_A \tag{4.14}$$

where $vp_A$ is the *vapor pressure* of the chemical $A$, and $X_A$ is the mole fraction of $A$ in the liquid phase. The mole fraction, $X_A$, can be related to liquid phase concentrations as follows:

$$X_A = \frac{C_A}{C} \qquad (4.15)$$

where $C_A$ is the molar concentration of $A$, and $C$ is the total molar concentration of the solution.

## 4.3.2.4 Henry's Law

Henry's Law states that the partial pressure of a chemical, $p_A$, in an air-water binary system at equilibrium is linearly proportional to its mole fraction in the aqueous phase, $X_A$, as long as the solution is dilute. The proportionality constant is known as Henry's Constant, $H$:

$$p_A = HX_A \qquad (4.16)$$

The above statement is conceptually the same as the partitioning phenomenon discussed in Section 4.3, where $K_{a-w}$ is comparable to $H$. The higher the value of $H$, the higher the tendency of the chemical to partition into the gas phase. Or, in other words, $H$ can be considered as a measure of the volatility of a chemical. As defined above, $H$ may take the dimensions of atm./mole fraction or mm Hg/mole fraction; similarly, $K_{a-w}$ can also take different forms. Table 4.2 summarizes the different forms of Henry's Law and conversion factors to relate them to one another.

***Worked Example 4.3***

The air-water partition coefficient, $K_{a-w}$, for oxygen has been reported as 40,000 atm.-mole/mole. (1) Estimate the dissolved oxygen concentration that can be expected in a natural body of pristine water. (2) Convert the given $K_{a-w}$ value to a molar concentration ratio form.

*Solution*

(1) The air-water partition coefficient discussed in Section 4.3 can be used to find the dissolved concentration, because the atmospheric content of oxygen is known as 21%. Consistent units have to be used in the calculations.

Air-water partition coefficient,

$$K_{a-w} = \frac{\text{Oxygen content in air}}{\text{Oxygen content in water}}$$

Hence,

$$\text{Oxygen content in water} = \frac{\text{Oxygen content in air}}{K_{a-w}}$$

Table 4.2  Different Forms of Quantifying Phase Contents and the Resulting Forms of Henry's Constant

| Gas Phase Content | Aqueous Phase Content | Form of Henry's Constant | Multiplication Factors for Converting to Other Forms | | | |
|---|---|---|---|---|---|---|
| | | | $H_{ppmf}$ | $H_{mcmc}$ | $H_{mfmf}$ | $H_{ppmc}$ |
| Partial pressure (atm.) | Mole fraction (mole/moles) | $H_{ppmf}$ [atm.] | 1 | $RT$ | $\rho RT/p$ | $\rho RT$ |
| Molar concentration (moles/L) | Molar concentration (moles/L) | $H_{mcmc}$ (–) | $1/RT$ | 1 | $p/\rho$ | $\rho$ |
| Mole fraction (moles/moles) | Mole fraction (moles/moles) | $H_{mfmf}$ (–) | $p/\rho RT$ | $\rho/p$ | 1 | $p$ |
| Partial pressure (atm.) | Molar concentration (moles/L) | $H_{ppmc}$ (atm.-L/mole) | $1/\rho RT$ | $1/p$ | $1/p$ | 1 |

$R$ = Ideal Gas constant; $T$ = absolute temperature; $\rho$ = molar density of water; $p$ = total pressure; typical units indicated as ( ).

The given value of $K_{a-w}$ indicates that the gas phase content is quantified in partial pressure (atm.), and the aqueous phase content is quantified by mole fraction (mole/mole). Oxygen content in the atmosphere = 21% = mole fraction of 0.21. Because the atmospheric pressure is 1 atm., using Dalton's Law, the partial pressure of oxygen in air = 0.21 × 1 atm. = 0.21 atm.

$$\therefore \text{Oxygen content in water} = \frac{0.21 \text{ atm.}}{40,000 \frac{\text{atm.-moles H}_2\text{O}}{\text{moles O}_2}}$$

$$= 5.25 \times 10^{-6} \frac{\text{moles O}_2}{\text{moles H}_2\text{O}}$$

$$= \left(5.25 \times 10^{-6} \frac{\text{moles O}_2}{\text{moles H}_2\text{O}}\right)\left(\frac{32 \text{ g O}_2}{\text{gmole O}_2}\right)\left(\frac{1000 \text{ mg}}{\text{g}}\right)\left(\frac{\text{moles H}_2\text{O}}{18 \text{ g H}_2\text{O}}\right)\left(\frac{1000 \text{ g}}{\text{L}}\right)$$

$$= 9.3 \frac{\text{mg}}{\text{L}}$$

(2) In the given $K_{a-w}$ value, the gas phase content is in partial pressure form, and the aqueous phase content is in the mole fraction form. To convert this value to the mole concentration ratio form, the gas phase content has to first be converted from the partial pressure form to the molar concentration form (moles/L). This can be achieved using the Ideal Gas Law:

$$pV = nRT$$

or,

$$\frac{n}{V} = \frac{p}{RT}$$

Assuming ambient temperature of $T = 25°C$, and $R = 82$ atm.-L/kmole-K, at the partial pressure = 0.21 atm.,

$$\frac{n}{V} = \frac{0.21 \text{ atm.}}{82 \frac{\text{atm.-L}}{\text{kmole-K}} (273 + 25)\text{K}} = 8.6 \times 10^{-6} \frac{\text{kmole}}{\text{L}} = 8.6 \times 10^{-3} \frac{\text{mole}}{\text{L}}$$

The mole fraction in the aqueous phase was found as $5.25 \times 10^{-6}$ in part (*1*), which can be converted to molar concentration $C$ using Equation (4.15):

$$C_A = X \times C = (5.25 \times 10^{-6}) \times \left(55.5 \frac{\text{gmole}}{\text{L}}\right) = 2.9 \frac{\text{gmole}}{\text{L}}$$

$$\text{Hence, } K_{a-w} = \frac{\dfrac{\text{Moles oxygen}}{\text{Volume of air}}}{\dfrac{\text{Moles oxygen}}{\text{Volume of water}}} = \frac{8.6 \times 10^{-3}}{2.9 \times 10^{-4}} = 29.6$$

Note that the oxygen content in the water is $\ll 0.02$ mole fraction, which satisfies the assumption of dilute solution, thus justifying the use of linear partitioning.

## 4.4 ENVIRONMENTAL TRANSPORT PROCESSES

Chemicals can be transported through the various compartments of the environment by microscopic level and macroscopic level processes. At the microscopic level, the primary mechanism of transport is by molecular *diffusion* driven by concentration gradients. At the macroscopic level, mixing (due to turbulence, eddy currents, velocity gradients) and bulk movement of the medium are the primary transport mechanisms. Transport by molecular diffusion and mixing has been referred to as *dispersive* transport, while transport by bulk movement of the medium is referred to as *advective* transport. Advective and dispersive transport are fluid-element driven, whereas diffusive transport is concentration-driven and can proceed under quiescent conditions. In this section, fundamentals of diffusive, dispersive, and advective transport mechanisms are reviewed along with the theories used to model the mass transfer phenomenon.

### 4.4.1 DIFFUSIVE TRANSPORT

Diffusive transport at the molecular level can take place under steady or unsteady conditions in homogeneous (gases, soils, water) or multiphase (sediments, biofilms) engineered and natural environmental systems. The rate of chemical transport under these conditions can be quantified using Fick's Laws of diffusion as summarized next.

#### 4.4.1.1 Steady State Conditions

The diffusive transport rate under *steady state conditions* can be quantified using Fick's First Law of diffusion. According to Fick's First Law, the molar rate of transport by diffusion in the $x$-direction, $J_{x,i}$ ($MT^{-1}$), is directly proportional to the concentration gradient, $dC_i/dx$ ($ML^{-3} - L^{-1}$), and the area of flow, $A_x$ ($L^2$):

$$J_{x,i} \propto A_x \left( \frac{dC_i}{dx} \right) \tag{4.17}$$

By introducing a proportionality constant, $D_i$, called the molecular diffusion coefficient ($L^2T^{-1}$), and a negative sign to indicate that the flux is positive in the $x$-direction,

$$J_{x,i} = -D_i A_x \left( \frac{dC_i}{dx} \right)$$  (4.18)

### 4.4.1.2 Unsteady State Conditions

The diffusive transport rate under time-dependent, *unsteady* state can be quantified using Fick's Second Law:

$$\frac{\partial}{\partial x} \left( -D_i \frac{\partial C_i}{\partial x} \right) = \frac{\partial C_i}{\partial t}$$  (4.19)

The above equations can be applied to diffusive transport through gases or liquids. The diffusion coefficient (or diffusivity), $D_i$, is an intrinsic molecular property for a chemical-solvent system. Tabulated numerical values for $D$ can be found in handbooks; they can also be estimated from chemical and thermodynamic properties following empirical correlations such as the Wilkie-Chang equation for diffusion of small molecules through water and the Chapman equation for diffusion in gases.

### 4.4.1.3 Multiphase Diffusion

In certain environmental systems, molecules may diffuse through a matrix of multiple phases. A typical example is the diffusion of chemical vapors through the vadose zone matrix that may consist of air, water vapor, pure chemical liquid, and soil. The effective diffusion coefficient under these conditions will be dependent upon the pore characteristics and can be accounted for by the *tortuosity* factor, $\tau$, to modify the pure phase diffusivity as follows:

$$D_{\text{pore},j} = D_{i,j} \left( \frac{\theta}{\tau} \right)$$  (4.20)

where $D_{\text{pore},j}$ is the diffusivity in the pores filled with phase $j$, $D_{i,j}$ is the molecular diffusivity in phase $j$, and $\theta$ is the porosity of the matrix.

***Worked Example 4.4***

The molecular diffusivity of nitrates in water is $19 \times 10^{-6}$ cm$^2$/s. In a river, nitrate concentration in the water column is 20 mg/L, and in the sediment pore waters, at a depth of 10 cm, it is 0.05 mg/L. Estimate the diffusive flux of nitrate into the sediments, assuming sediment bed porosity of 65% and a tortuosity factor of 3.

*Solution*

The flux, $N$, which is the diffusive mass flow rate, $J$, per unit area, $A$, can be found from Fick's First Law and from using diffusivity in the pore waters. Diffusivity in the pore waters can be found using the porosity and tortuosity factors.

The flux can be calculated from the following:

$$N = \frac{J}{A} \cong D\frac{\Delta C}{\Delta z} = [D_{pore.w}]\frac{(C_w - C_s)}{\Delta z} = \left[D_w\left(\frac{\theta}{\tau}\right)\right]\frac{(C_w - C_s)}{\Delta z}$$

where,

$D_W = 19 \times 10^{-6}$ cm$^2$/s, $\theta = 0.65$, $\tau = 3$, $C_W = 20$ mg/L, $C_s = 0.05$ mg/L, and $\Delta z = 10$ cm

Hence, nitrate flux into the sediment

$$= \left\{\left(19 \times 10^{-6} \frac{cm^2}{s}\right)\left(\frac{0.65}{3}\right)\frac{(20 - 0.05)\frac{mg}{L}}{10\ cm}\right\}\left(\frac{3600\ s}{hr}\right)\left(\frac{L}{1000\ cm^3}\right)\left(\frac{10^4\ cm^2}{m^2}\right)$$

$$= 3.0 \times 10^{-4} \frac{g}{m^2 - hr}$$

## 4.4.2 DISPERSIVE TRANSPORT

Dispersive transport results from a combination of multiple mechanisms, such as molecular diffusion, turbulence, eddy currents, and velocity gradients. The exchange of momentum between fluid elements in a turbulent flow field is the driving force for this mode of transport. The quantification of concentration profiles and chemical fluxes by dispersive transport follows the same model as that for diffusive transport (discussed in Section 4.4.1) but uses a dispersion coefficient, $E$ ($L^2T^{-1}$).

## 4.4.3 ADVECTIVE TRANSPORT

Advection is the mechanism by which a chemical is transported across the boundary of the system and through the system by the flow of the bulk medium. The molar flux of chemical $i$ transported by advection in the $x$-direction, $N_{x,i}$ ($ML^{-2}T^{-1}$), can be found from

$$N_{x,i} = v_x C_i \qquad (4.21)$$

where $v_x$ is the velocity of flow ($LT^{-1}$) in the $x$-direction and $C_i$ is the molar concentration of chemical $i$ in the bulk medium. If $A_x$ ($L^2$) is the area normal to the flow, the molar transport rate, $J_{x,i}$ ($MT^{-1}$) by advection is therefore found from the following:

$$J_{x,i} = A_x N_{x,i} = A_x v_x C_i = Q C_i \tag{4.22}$$

where $Q$ is the volumetric flow rate ($L^3 T^{-1}$) of the bulk medium.

## 4.5 INTERPHASE MASS TRANSPORT

The above sections dealt with intraphase transport processes. The transfer of mass from one phase to another is an important transport process in environmental systems. Examples of such processes include aeration, reaeration, air stripping, soil emissions, etc. The following theories have been proposed to model such transfers: the Two-Film Theory proposed in the 1920s, the Penetration Theory proposed in the 1930s, and the Surface Renewal Theory proposed in the 1950s. Among these, the first is the simplest, most understood, and most commonly used. As such, only the Two-Film Theory is reviewed here.

### 4.5.1 TWO-FILM THEORY

The Two-Film Theory can best be illustrated using the classical example of transfer of oxygen in an air-water binary system as shown in Figure 4.3. According to this theory, the following are postulated:

- There are two films at the interface, one on each side.
- Concentration gradients exist only within the two films, and the bulk are well mixed.
- Concentrations, $C_{a,i}$ and $C_{w,i}$, at the interface are at equilibrium.

Because the interfacial concentrations are at equilibrium, using linear partitioning

$$\frac{C_{a,i}}{C_{w,i}} = K_{a-w} \tag{4.23}$$

The molar flow rate through the gas-side film, $J_{x,i,a}$, is given by

$$J_{x,i,a} = D_{i,a} A_x \left( \frac{dC_i}{dx} \right) = \frac{D_{i,a} A_x (C_{a,b} - C_{a,i})}{\Delta x_a} = k_g A_x (C_{a,b} - C_{a,i}) \tag{4.24}$$

and, the molar flow rate through the liquid-side film, $J_{x,i,w}$, is given by

$$J_{x,i,w} = D_{i,w} A_x \left( \frac{dC_i}{dx} \right) = D_{i,w} A_x \frac{(C_{w,i} - C_{w,b})}{\Delta x_w} = k_l A_x (C_{w,i} - C_{w,b}) \tag{4.25}$$

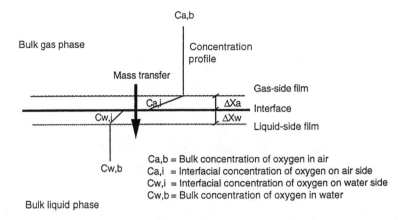

**Figure 4.3** Illustration of the Two-Film Theory.

where $k_g = D_{i,a}/\Delta x_a$ and $k_l = D_{i,w}/\Delta x_w$ are *local mass transfer coefficients* $(LT^{-1})$ for the gas- and liquid-side films, respectively. Under steady state conditions, the above two expressions for the molar flow rates must equal one another: $J_{x,i,a} = J_{x,i,w} = J_{x,i}$.

Because the interfacial concentrations are not known, a new variable is introduced to make the above equations useful. The new variable is defined as the liquid phase concentration, $C_w^*$, that would be in equilibrium with the current gas phase concentration; in other words, $C_{a,b} = K_{a-w}C_w^*$. Thus, equating the two molar flow rate equations and eliminating $C_{a,b}$ and $C_{a,i}$, an expression for $C_{w,i}$ can be found as follows:

$$C_{w,i} = \frac{k_l C_{w,b} + k_g K_{a-w} C_w^*}{k_l + K_{a-w} k_g} \tag{4.26}$$

On substituting the above result back into the expression for the flux through the liquid film, the mass transfer flux is finally found as follows:

$$N_{x,i} = \frac{J_{x,i}}{A_x} = K_L(C_w^* - C_{w,b}) \tag{4.27}$$

where the new coefficient $K_L$ $(LT^{-1})$ is known as the *overall mass transfer coefficient* relative to the liquid, which is related to the local mass transfer coefficients by

$$K_L = \frac{1}{\dfrac{1}{k_g K_{a-w}} + \dfrac{1}{k_l}} \tag{4.28}$$

The mass transfer rate per unit volume of the bulk medium can then be found by multiplying the flux by the specific interfacial mass transfer area, $a$. In most systems, this may be not readily accessible and will have to be estimated. In addition, the overall mass transfer coefficients are highly system specific and also have to be determined experimentally. In practice, the combined term $K_L a$ is estimated using empirical correlations developed from similar systems. A compilation of correlations for estimating $K_L a$ in environmental systems can be found in Webber and DiGiano (1996).

## 4.6 ENVIRONMENTAL NONREACTIVE PROCESSES

As defined earlier, nonreactive processes cause changes in the chemical content within the system, without the chemical undergoing any change in its molecular structure; the chemical may, however, undergo a phase change. Such processes are also referred to as physical processes. Some examples of environmental nonreactive processes are settling, resuspension, flotation, adsorption, desorption, absorption, thermal desorption, volatilization, extraction, filtration, membrane processes, and biosorption. A review of some of the more common nonreactive processes is included here.

### 4.6.1 ADSORPTION AND DESORPTION

Adsorption and desorption of chemicals (adsorbates) at liquid-solid and gas-solid interfaces (adsorbents) are ubiquitous in natural and engineered systems. Examples include adsorption of molecules onto sediments, suspended matter, soil, and aerosols in natural systems and onto activated carbon, zeolite, and ion exchange resins in engineered systems. Two types of mechanisms are thought to be significant in these processes: physisorption and chemisorption. Physisorption is driven by van der Waals, electrostatic, and hydrophobic forces. Chemisorption is driven by covalent bonding forces. In practice, both of these mechanisms often occur together, and a generalized approach is used to model the process.

The relationship between the concentration of the adsorbate on the adsorbent (solid) and in the bulk phase (gas or liquid) is often referred to as the *isotherm*. Two of the more common isotherms used for aqueous systems are as follows:

$$\text{Langmuir isotherm: } C_s = \frac{AC_w}{B + C_w} \tag{4.29}$$

$$\text{Freundlich isotherm: } C_s = K_f C_w^{1/n} \tag{4.30}$$

where $C_s$ is the concentration of the adsorbate on the adsorbent $(MM^{-1})$; $C_w$ is the concentration of the adsorbent in the bulk liquid phase $(ML^{-3})$; and $A$, $B$, $K_f$, and $n$ are empirical constants to be determined by experimentation. The model constants can also be estimated from the molecular structures of the adsorbates using quantitative structure activity relationships (QSARs). As can be noted, the above isotherms are nonlinear and can result in nonlinear models that may be difficult to solve by traditional mathematical calculi.

For most environmental systems, where the maximum adsorptive capacity of the adsorbent is much greater than the actual adsorbed concentration, the index $n$ in Freundlich isotherm may be assumed to be approximately equal to 1. Thus, the Freundlich isotherm can be reduced to a linear model, where the other constant, $K_f$, while being similar to the partition coefficient discussed in Section 4.3, now has the dimensions of $L^3M^{-1}$:

$$\text{Linear isotherm: } C_s = K_f C_w \qquad (4.31)$$

Usually, adsorption/desorption processes are very fast, attaining equilibrium condition within minutes or hours. This implies that in environmental systems where the time step in calculations is in the order of days, the assumption of equilibrium can be justified. However, if it is necessary to include the rate expression, as in microscopic analysis, a first-order expression can be assumed to give the rate of transfer:

$$\frac{dC_s}{dt} = k(C_s^* - C_s) \qquad (4.32)$$

where $C_s^*$ is the equilibrium concentration on the solid phase.

## 4.6.2 SETTLING AND RESUSPENSION

The fate and transport of suspended solids and of chemicals adsorbing onto them or desorbing from them can be impacted by settling or resuspension processes. Examples include sedimentation of suspended particles in water and wastewater treatment, electrostatic precipitation in air pollution control, and settling of algae in natural systems. These processes are driven by gravitational, buoyancy, hydrodynamic, aerodynamic, or electrostatic forces. The two processes can be combined and modeled as a net settling process.

A submodel for the removal rate of suspended particles from the bulk fluid can be developed as follows, considering a well-mixed lake as an example. The rate of change of mass, $m$, of suspended particles in the water column should equal the rate of settling of suspended particles toward the sediment:

$$\text{Rate of change of suspended particles} = \frac{dm}{dt} = \frac{d(P_w V)}{dt} \qquad (4.33)$$

$$\text{Rate of settling of suspended particles to sediement} = A v_s P_w$$

where $P_w$ is the concentration of the particles in the water column ($ML^{-3}$), $V$ is the volume ($L^3$) of the lake, $A$ is the average surface area of the lake ($L^2$), and $v_s$ is the settling velocity of the particles ($LT^{-1}$). Hence, equating the above two,

$$\frac{dm}{dt} = -Av_s P_w = -Av_s \left(\frac{m}{V}\right) = -\left(\frac{v_s}{D}\right)m = -k_s m \qquad (4.34)$$

where $D$ is the depth (L) of the lake. From the above, it can be seen that the rate of settling is a first-order process, with a rate constant of $k_s = (v_s/D)$, having a dimension of $T^{-1}$. The settling velocity of the particles can be determined experimentally or theoretically using Stokes' Law or empirically using correlations. A similar submodel can be used in all the examples of settling processes mentioned earlier.

Once the settling rate is established, the sediment buildup rate can be deduced. Also, by combining the settling submodel with the adsorption submodel, the rate of removal of chemical from the water by adsorption by suspended solids, and hence the removal of chemicals from the water column, and finally, its accumulation in the sediments can be modeled.

The relationship between adsorbed concentration, $C_s$, particulate concentration, $P_w$, dissolved concentration, $C_w$, and the adsorbed bulk concentration, $C_P$, can be developed from the following definitions:

$$C_s = \frac{\text{Mass of adsorbed chemical}}{\text{Mass of suspended particles}} \qquad (4.35)$$

$$P_w = \frac{\text{Mass of suspended particles}}{\text{Volume of water}} \qquad (4.36)$$

$$C_w = \frac{\text{Mass of dissolved chemical}}{\text{Volume of water}} \qquad (4.37)$$

$$C_P = \frac{\text{Mass of adsorbed chemical}}{\text{Volume of water}} \qquad (4.38)$$

Hence, assuming linear adsorption isotherm,

$$C_P = C_s P_w = (K_f C_w)P_w$$

$$\text{Since } C_T = C_P + C_w \qquad (4.39)$$

where $C_T$ is the total concentration of the chemical in the water column, the fractions of the dissolved and particulate forms of the chemical can be found from:

$$\text{Fraction in dissolved form} = f_d = \frac{C_w}{C_T} = \frac{1}{(1 + K_f P_w)} \qquad (4.40)$$

$$\text{Fraction in particulate form} = 1 - f_d = \frac{K_f P_w}{(1 + K_f P_w)} \qquad (4.41)$$

**Worked Example 4.5**

A lake has been receiving a constant input of suspended solids and a toxicant, resulting in a suspended solids concentration of 10 mg/L in the lake. The partitioning of the toxicant in a solids-water system has been found to be linear with a partition coefficient of 200,000 L/kg. The net settling velocity of the suspended solids is 0.6 ft/day, the average depth of the lake is 30 ft, and the volume of the lake is $3 \times 10^6$ cu ft. Determine the following: (1) adsorbed concentration, (2) the rate constant for the settling process, (3) the dissolved and particulate fractions of the toxicant in the water column, (4) the flux of solids to the sediment, and (5) the flux of toxicant to the sediment if the total concentration of the toxicant in the water column is 10 μg/L.

*Solution*

(1) Using Equation (4.31), adsorbed concentration, $C_s$

$$= K_f \bullet C_w = (200{,}000 \text{ L/kg}) * (10 \text{ mg/L}) = 2 * 10^6 \text{ mg/kg}$$

(2) Using Equation (4.34), the rate constant, $k_s$

$$= v_s/D = (0.6 \text{ ft/day})/30 \text{ ft} = 0.02 \text{ day}^{-1}$$

(3) Using Equation (4.40), the dissolved fraction, $f_d$

$$= \frac{1}{1 + K_f P_w} = \frac{1}{1 + 200{,}000 \frac{L}{kg} \times 10 \frac{mg}{L} \times \left(10^{-6} \frac{kg}{mg}\right)} = \frac{1}{1 + 2} = 0.67$$

and, the particulate fraction $f_p = 1 - f_d = 1 - 0.67 = 0.33$

(4) The rate of solids settling to the sediments

$$= \frac{dm}{dt} = k_s m = k_s (V P_w)$$

$$= \left(0.02 \frac{1}{day}\right)\left[\left(3 \times 10^6 \text{ ft}^3 \times \left\{\frac{1}{28.32 \text{ ft}^3}\right\}\right)\left(10 \frac{mg}{L} \times \left\{\frac{1 \text{ g}}{1000 \text{ mg}}\right\}\right)\right] = 21.2 \frac{g}{day}$$

The flux of solids to the sediment

$$= \frac{\left(\dfrac{dm}{dt}\right)}{A} = \frac{\left(\dfrac{dm}{dt}\right)}{\left(\dfrac{V}{D}\right)} = \frac{\left(21.2\ \dfrac{\text{g}}{\text{day}}\left[\dfrac{1000\ \text{mg}}{\text{g}}\right]\right)}{\left(\dfrac{3 \times 10^6\ \text{ft}^3}{30\ \text{ft}}\right)} = 0.2\ \frac{\text{mg}}{\text{ft}^2 - \text{day}}$$

(5) The flux of toxicant to the sediments

= Flux of solids * Adsorbed concentration

First, find dissolved concentration in the water column:

$$C_w = f_d C_T = (0.67)\left(10\ \frac{\mu\text{g}}{\text{L}}\right) = 6.7\ \frac{\mu\text{g}}{\text{L}}$$

Now, find adsorbed concentration using Equation (4.31)

$$C_s = K_f C_w = \left(200,000\ \frac{\text{L}}{\text{Kg}}\right)\left(6.7\ \frac{\mu\text{g}}{\text{L}}\right) = 1.34 \times 10^6\ \frac{\mu\text{g}}{\text{kg}}$$

Hence, the flux of toxicants to sediments

$$= \left(0.2\ \frac{\text{mg}}{\text{ft}^2 - \text{day}}\right) \times C_s = \left(0.2\ \frac{\text{mg}}{\text{ft}^2 - \text{day}}\right)\left(1.34 \times \left[\frac{10^6\ \text{mg}}{\text{g}}\right] \times \left[\frac{1\ \text{kg}}{10^9\ \mu\text{g}}\right]\right)$$

$$= 0.268 \times 10^{-3} = 0.268 \times 10^{-3}\ \frac{\text{mg}}{\text{ft}^2 - \text{day}} = 0.268\ \frac{\mu\text{g}}{\text{ft}^2 - \text{day}}$$

## 4.6.3 VOLATILIZATION AND ABSORPTION

Volatilization and absorption involve phase change and are commonly encountered processes in engineered and natural systems. Examples include aeration, air stripping, reaeration, evaporation, soil venting, and emissions. These processes are driven by concentration gradients and can be modeled using the Two-Film Theory as discussed in Section 4.5.1.

For example, the emission of volatile organic chemicals (VOCs) from an aeration tank fitted with a surface aerator to the atmosphere by volatilization can be described by the following:

$$\frac{dC_w}{dt} = K_L a(C_w - C_w^*) = K_L a\left(C_w - \frac{C_a}{H}\right) \approx K_L a C_w \qquad (4.42)$$

where $K_L a$ is the overall mass transfer coefficient, $(\text{T}^{-1})$, $C_w^*$ is the concentration of the VOC in water that would be in equilibrium with the concentration in the air $(M/L^{-3})$, $C_a$ is the concentration of the VOC in air $(M/L^{-3})$, and $H$ is the Henry's Constant $(-)$ of the VOC. Because the VOC concentration in the atmosphere can be assumed to be negligible, the equation reduces to a first-order process with respect to $C_w$. The mass transfer coefficient is often found from empirical correlations specific to the system.

### 4.6.4 BIOUPTAKE

Chemicals and biota in water interact through several different processes, such as biosorption, biouptake, excretion, depuration, and biodegradation. These processes result in bioconcentration, bioaccumulation, and through the food chain, in biomagnification. Bioconcentration occurs through uptake from the dissolved phase, mainly through a partitioning process. Bioaccumulation results from uptake from water, ingestion of suspended solids that carry adsorbed chemicals, and prey. Biomagnification is the increase in body burden through each step of the trophic ladder.

The rate of increase of chemical in biomass can be expressed as follows:

$$\frac{dF}{dt} = \eta k_1 C_w - k_2 F \tag{4.43}$$

where $\eta$ is the efficiency of absorption by the gills $(-)$, $k_1$ is the rate constant $(L^3 M^{-1} T^{-1})$, $k_2$ is the depuration rate constant $(T^{-1})$, and $F$ is the concentration of the chemical in the biomass $(MM^{-1})$. At steady state, the above reduces to

$$F = \left(\frac{\eta k_1}{k_2}\right) C_w = BCF C_w \tag{4.44}$$

where $BCF$ is termed the bioconcentration factor $(L^3 M^{-1})$. Notice that $BCF$ is very similar to the adsorption constant, $K_f$, for linear isotherms.

### Worked Example 4.6

The bioconcentration of a new pesticide in a lake is to be evaluated. The suspended solids concentration in the lake = 20 mg/L, and the water-solids partition coefficient of the pesticide is estimated as 10,000 L/kg. Assuming the $BCF$ of the pesticide as 10,000 L/kg, estimate its total concentration in the water column that can be tolerated without violating the Food and Drug Administration level of 0.3 ppm in the fish.

### Solution

Using Equation (4.44), the dissolved concentration, $C_w$, that can be tolerated

$$= \frac{F}{BCF} = \frac{0.3 \frac{g}{10^6 \, g}}{5000 \frac{L}{kg}\left[\frac{kg}{10^9 \mu g}\right]} = 0.06 \frac{\mu g}{L}$$

Using Equation (4.40), the dissolved fraction, $f_d$, is given by:

$$f_d = \frac{1}{1 + K_f P_w} = \frac{1}{1 + \left(10,000 \frac{L}{kg}\right)\left(20 \frac{mg}{L}\right)\left(\frac{1 \, kg}{10^6 \, mg}\right)} = 0.83$$

Hence, the $C_T$ allowable in the water column $= C_w/f_d = 0.06/0.83 = 0.07$ µg/L.

## 4.7 ENVIRONMENTAL REACTIVE PROCESSES

Reactive processes result in changes in concentrations of chemicals within a system by degrading or transforming them to different chemical(s). Reactions can occur in a single step (elementary reactions) or in multiple steps, consecutively, in parallel, or in cycles. Some of the most common environmental reactive processes are biodegradation, hydrolysis, oxidation, reduction, incineration, precipitation, ion exchange, and photolysis. Reactive processes can be categorized as *homogeneous* if they occur only in a single phase or *heterogeneous* if they involve multiple phases.

Regardless of whether a reaction is homogeneous or heterogeneous, it is essential to quantify the rate at which chemicals are degraded or transformed by that reaction in order to model the chemicals' fate and transport. When the rates of reactions are faster than transport processes (discussed in Section 4.4), it is reasonable to assume that the system is at *chemical equilibrium,* allowing concentrations of participating chemical species to be established through *stoichiometry.* Acid-base and complexation reactions are examples of fast reactions, while redox reactions are typically slower. When the reaction rates are comparable to transport rates, concentrations of participating chemical species have to be determined through reaction kinetics. In the next sections, the fundamentals of stoichiometry and kinetics of some of the more common environmental reactions are reviewed.

### 4.7.1 CHEMICAL EQUILIBRIUM

Chemical equilibrium is said to exist when the forward rate of a reversible reaction is equal to the backward rate. The concept of chemical equilibrium is best illustrated by considering a simple case of a reversible reaction where species A and B are participating: $A \rightleftharpoons B$. The rates of the forward and backward reactions can expressed in terms of their respective reaction rate constants and the molar concentrations of the species by:

$$r_{\text{forward}} = k_1[A] \quad \text{and} \quad r_{\text{backward}} = k_2[B] \tag{4.45}$$

Therefore, at equilibrium, equating the two rates, the concentrations of the two species at equilibrium are related by:

$$\frac{[B]_{\text{eq}}}{[A]_{\text{eq}}} = \frac{k_1}{k_2} = K_{\text{eq}} \tag{4.46}$$

where $K_{\text{eq}}$ is known as the *equilibrium constant* for the reaction. Equilibrium constants play an important role in environmental modeling in that they are key inputs in determining equilibrium distributions of participating reactants and products in many environmental systems. It has to be noted that $K_{\text{eq}}$ is strongly dependent on the system temperature, for example, in the case of self-ionization of water, $K_{\text{eq}} = 0.45 \times 10^{-14}$ at 15°C and $K_{\text{eq}} = 1.47 \times 10^{-14}$ at 30°C.

The directions of reversible reactions depend on the energy of the system and can be established through a thermodynamic analysis. It can be shown that $K_{\text{eq}}$ is related to the Gibbs' standard free energy of the products minus the Gibbs' standard free energy of the reactants, $\Delta G^\circ$, by the following equation:

$$\Delta G^\circ = -RT \ln K_{\text{eq}} \tag{4.47}$$

where $R$ is the Ideal Gas Constant and $T$ is the absolute temperature. This equation enables $K_{\text{eq}}$ values to be calculated from $\Delta G^\circ$. Tabulated values of $K_{\text{eq}}$ can also be found in Chemical Property Handbooks. The application of the above concepts is illustrated in Worked Example 4.7.

**Worked Example 4.7**

(1) Calculate the equilibrium constant for the dissociation of carbonic acid.
(2) Derive the equations that can be used to plot the concentrations of the various carbonate species in a closed aqueous system as a function of pH for a given total carbon mass, $C_T$.

*Solution*

(1) The equilibrium constant can be calculated from $\Delta G^\circ$ using Equation (4.47). By definition, $\Delta G^\circ$ can be found from the following:

$$\Delta G^\circ = \sum_i (v_i \Delta G_f^\circ)_{\text{products}} - \sum_j (v_j \Delta G_f^\circ)_{\text{reactants}}$$

where $v_i$ and $v_j$ are the stoichiometric coefficients in the equation, and $\Delta G_f^o$ are the standard free energy of each reactant and product. The relevant equation in this example is as follows:

$$H_2CO_3^* \leftrightarrows H^+ + HCO_3^-$$

Looking up the $\Delta G_f^o$ values and substituting,

$$\Delta G^o = -586.8 + 0 - (-623.2) = 36.5 \frac{kJ}{mole}$$

Hence,

$$K = \exp\left(\frac{-\Delta G^o}{RT}\right) = \exp\left(\frac{-36.5 \frac{kJ}{mole}}{0.008314 \frac{kJ}{mole\text{-}K} (25 + 273)K}\right) = 4.04 \times 10^{-7}$$

(2) The relevant equations are first compiled along with their respective equilibrium constants:

$$H_2CO_3^* \leftrightarrows HCO_3^- + H^+ \qquad K_1 = \frac{[H^+][CO_3^-]}{[H_2CO_3]} = 10^{-6.3}$$

$$HCO_3^- \leftrightarrows CO_3^{2-} + H^+ \qquad K_2 = \frac{[H^+][CO_3^{2-}]}{[HCO_3^-]} = 10^{-10.2}$$

$$H_2O \leftrightarrows OH^- + H^+ \qquad K_w = \frac{[H^+][OH^-]}{[H_2O]} = 10^{-14}$$

The above expressions can be formulated as a set of simultaneous equations:

$$\log K_1 = \log [H^+] + \log [HCO_3^-] - \log [H_2CO_3]$$

$$\log K_2 = \log [H^+] + \log [CO_3^{2-}] - \log [HCO_3^-]$$

$$\log K_w = \log [H^+] + \log [OH^-]$$

In addition to the above equations, a total mass balance on carbon can be written:

$$C_T = [H_2CO_3] + [HCO_3^-] + [CO_3^{2-}]$$

The above set of linear simultaneous equations has to be solved to develop the required plot. See Modeling Example 9.6 in Chapter 9 for a computer implementation of the solution process.

## 4.7.2 ELEMENTARY REACTIONS

The rate or *kinetics* of these reactions can be quantified using the Law of Mass Action, which states that the rate is proportional to the concentration of the reactants. For example, consider the reaction: A + B → C. The rate of consumption of species A by this reaction is given by the Law of Mass Action as follows:

$$\frac{dC_A}{dt} = -kf(C_A, C_B) \tag{4.48}$$

where $k$ is a temperature-dependent reaction rate constant, and the function $f(C_A, C_B)$ has to be determined experimentally. A common general form of this function is

$$\frac{dC_A}{dt} = -kC_A^{\alpha}, C_B^{\beta} \tag{4.49}$$

where the powers to which the concentrations are raised are referred to as the *reaction order*. In the above example, the reaction is of order $\alpha$ with respect to species A, and order $\beta$ with respect to B, and the overall order of the reaction is $n = (\alpha + \beta)$. Because most modeling efforts build upon submodels of single chemicals, the following sections will deal with the equation:

$$\frac{dC}{dt} = -kC^n \tag{4.50}$$

where $n = 0$ for zero order, $n = 1$ for first order, or $n = 2$ for second order. The mass removal rate by such a reaction can be found from $V(dC/dt)$, where $V$ is the volume in which the reaction is occurring. The change in concentration as a function of time and the time to reduce the initial concentration by 50% (*half-life*) can be found by solving the above equation for the appropriate value of $n$ and initial concentration of $C_0$. The results are summarized in Table 4.3. The order of the reaction and the rate constant have to be determined experimentally. The procedures are well established and can be found in several textbooks on reaction kinetics.

## 4.7.3 ENZYME-MEDIATED REACTIONS

The Michaelis-Menten model of microbial growth is a classic example of an enzyme-mediated two-step reaction where the substrate $S$ is converted to the product $P$ in the presence of a catalytic enzyme, $E$:

$$E + S \leftrightarrows [ES] \rightarrow E + P$$

Table 4.3 Reactions of Order Zero, First, and Second

| | Order of Reaction | | |
| --- | --- | --- | --- |
| | Zero | First | Second |
| Value of $n$ | $n = 0$ | $n = 1$ | $n = 2$ |
| Governing equation: | $\dfrac{dC}{dt} = -k$ | $\dfrac{dC}{dt} = -kC$ | $\dfrac{dC}{dt} = -kC^2$ |
| Mass removal rate: | $Vk$ | $VkC$ | $VkC^2$ |
| Concentration profile: | $C = C_0 - kt$ | $C = C_0 e^{-kt}$ | $C = \dfrac{C_0}{1 + kC_0 t}$ |
| Units of $k$: | $ML^{-3}T^{-1}$ | $T^{-1}$ | $L^3 M^{-1} T^{-1}$ |
| Half-life | $C_0/(2k)$ | $1/(1.44k)$ | $1/(kC_0)$ |

This model results in a rather complicated rate expression and does not yield a simple rate expression or reaction order:

$$\frac{dP}{dt} = \mu_{max} \frac{[P][S]}{K_M + [S]} \tag{4.51}$$

where $\mu_{max}$ is the maximum growth rate of the products. This expression is neither first order nor second order. At low substrate concentrations, $[S] \ll K_M$, the expression can be approximated as follows:

$$\frac{dP}{dt} = \mu_{max}[P][S] \tag{4.52}$$

to be second order overall; and, at high substrate concentrations, $[S] \ll K_M$, can be approximated to be first order in $P$ as follows:

$$\frac{dP}{dt} = \mu_{max}[P] \tag{4.53}$$

Thus, the results summarized in the previous section can be applied to complex rate equations such as the Michaelis-Menton model.

## 4.7.4 PHOTOLYSIS

This is a solar energy driven decay process by which chemicals may be transformed at a molecular level. This process can take place in two steps: direct photolysis due to sunlight and sensitized indirect photolysis due to excess energy. The direct photolysis process is well understood and can be modeled by a first-order process:

$$\frac{dP}{dt} = -k_p C \tag{4.54}$$

where $k_p$ is the first-order photolysis rate constant $(T^{-1})$. This rate constant depends on the chemical as well as the light intensity and has to be determined experimentally or estimated from correlations. In the case of lakes, for example, a depth-averaged $k_{pa}$ value can be found in terms of near-surface photolysis rate and light intensity:

$$k_{p,a} = k_{p,s} \left( \frac{I_0}{I_0^1} \right) \frac{RDF}{RDF_0} \left\{ \frac{1 - \exp(-K_e \lambda_{max} D)}{K \lambda_{max} D} \right\} \tag{4.55}$$

where $k_{p,s}$ is the measured, near-surface direct photolysis rate $(T^{-1})$, $I_1^0$ is the light intensity at which $k_{p,s}$ was measured, $RDF$ is the radiance distribution factor = 1.2–1.6, $RDF_0$ is the radiance distribution factor at the surface = 1.2, $k_e \lambda_{max}$ is the light attenuation coefficient, and $D$ is the depth (L).

## 4.7.5 HYDROLYSIS

The reaction of certain chemicals with water results in the cleavage of chemical bonds. The breakdown of a chemical by this process depends on molar concentrations of $[H^+]$ and $[OH^-]$ and can be described by the following:

$$\frac{dC_w}{dt} = -k_a[H^+]C_w - k_n C_w - k_b[OH^-]C_w \tag{4.56}$$

where $k_a$ $(M^{-1}T^{-1})$, $k_n$ $(T^{-1})$, and $k_b$ $(M^{-1}T^{-1})$ are the acid-, neutral-, and base-catalyzed hydrolysis reaction rate constants, respectively. Often, the above is approximated by a pseudo first-order reaction at a given pH according to:

$$\frac{dC_w}{dt} = -k_h C_w \tag{4.57}$$

using a first-order hydrolysis reaction rate constant, $k_h$ $(T^{-1})$.

## 4.7.6 BIOTRANSFORMATION

Microbially mediated transformation processes are common in a wide range of engineered and natural systems. These processes can include *mineralization*, in which organic compounds are converted to inorganic compounds; *detoxification*, in which toxic chemicals are transformed to innocuous byproducts; and *cometabolism*, in which compounds are used as secondary substrates and not as nutrients.

When the microbial population has adapted to the contaminant, the rate constant, $k_b$, for the transformation process can be modeled by the Michaelis-Menton equation:

$$k_b = \frac{\mu_{max}X}{Y(K_s + C_w)} \tag{4.58}$$

where $\mu_{max}$ is the maximum growth rate of the organisms $(T^{-1})$, $X$ is the biomass concentration $(ML^{-3})$, $Y$ is the yield coefficient $(-)$, $K_s$ is the half-saturation constant $(ML^{-3})$, and $C_w$ is the dissolved concentration of the substrate $(ML^{-3})$.

Because the above equation is nonlinear, it does not lend itself easily for continuous modeling. Often, in environmental systems, $C_w \ll K_s$; under such conditions, Equation (4.58) can be simplified to the following linear form:

$$k_b = \frac{\mu_{max}X}{YK_s} = k_{b2}X \tag{4.59}$$

Thus, the rate equation for the disappearance of the chemical by biotransformation simplifies to that of a second-order process, with a rate constant, $k_{b2}$ $(L^3M^{-1}T^{-1})$. Further, if the biomass concentration, $X$, is assumed to be a constant, the equation simplifies to that of a first-order process.

### Worked Example 4.8

The second-order decay rate of di-$n$-butyl has been reported as $7 \times 10^{-7}$ mL/cell-day. Estimate its half-life assuming a biomass of (1) 100 cells/mL and (2) 5000 cells/mL.

*Solution*

Half-life for first-order process = $1/1.44k$ where, in this case, $k = (k_{b2})X$

(1) When $X = 100$ cells/L, half-life

$$= \frac{1}{1.44\left(7 \times 10^{-7} \, \frac{mL}{cell\text{-}day}\right)\left(100 \, \frac{cells}{mL}\right)} = 9900 \text{ days}$$

(2) When $X = 5000$ cells/L, half-life

$$= \frac{1}{1.44\left(7 \times 10^{-7} \, \frac{mL}{cell\text{-}day}\right)\left(5000 \, \frac{cells}{mL}\right)} = 200 \text{ days}$$

## 4.8 MATERIAL BALANCE

As indicated in Chapter 2, the material balance approach is the primary principle upon which mathematical models of environmental systems are built. It is based on the principle of conservation of mass—*mass can neither be created nor destroyed.* Before applying the material balance principle, the following preliminaries have to be addressed:

- Define and characterize the system and the boundary.
- Identify the *inflows* and *outflows* of the target chemical crossing the boundary.
- Identify all *sources* and *sinks* of the target chemical inside the system.
- Identify all *reactive* and *nonreactive* processes acting upon the target chemical.

In the verbal form, the material balance (MB) equation can be stated as follows:

Net rate of change      Rate of inflow      Rate of outflow      Rate of
of material within = of material       − of material       ± reactions
system                        into system           from system          within system

The MB equation can be applied to distributed systems and lumped systems.

### 4.8.1 MATERIAL BALANCE FOR DISTRIBUTED SYSTEMS

In distributed systems, the material balance is applied in the differential form to a small element of the system. This procedure has already been detailed in Section 2.3 in Chapter 2.

### 4.8.2 MATERIAL BALANCE FOR LUMPED SYSTEMS

Key characteristics of the lumped system are that the system is homogeneous or well mixed and the concentration of the target chemical in the outflow stream(s) is the same as that within the system. An example of a lumped system is shown in Figure 4.4, where the system, boundary, two advective inflows, and two advective outflows are indicated.

The terms in the MB equation can be expressed as follows in terms of the system variables, assuming first-order reactions:

$$\text{Net rate of change of material within system} = \frac{d(VC)}{dt} \qquad (4.60)$$

$$\text{Rate of inflow of material into system} = \sum(QC)_{in} = Q_1C_1 + Q_2C_2 \quad (4.61)$$

Rate of outflow of material out of system= $\left[\sum(Q)_{\text{out}}\right]C = (Q_3 + Q_4)C$ (4.62)

$$\text{Rate of reactions within system} = CV\sum k \qquad (4.63)$$

Thus, the MB equation for this example can be expressed as follows:

$$\frac{d(CV)}{dt} = \sum(QC)_{\text{in}} - \sum(QC)_{\text{out}} \pm CV\sum(k) \qquad (4.64)$$

$$= Q_1C_1 + Q_2C_2 - (Q_3 + Q_4)C \pm CV\sum k \qquad (4.65)$$

The above general equation may be simplified by assuming the volume of the system, $V$, to be a constant.

***Worked Example 4.9***

An experimental method known as the equilibrium partitioning in closed systems (EPICS) method has been proposed by Lincoff and Gossett (1984) to measure the air-water partition coefficient (Henry's Constant) of volatile organic chemicals. In this method, a certain mass of the test chemical is injected into a sealed bottle containing known volumes of water, $V_{l,1}$, and air, $V_{g,1}$. The bottle is allowed to reach equilibrium, and the gas phase concentration, $C_{g1}$, is measured analytically. The experiment is repeated by injecting the same mass of the chemical into another bottle with different volumes of water, $V_{l,2}$, and air, $V_{g,2}$. Again, the bottle is allowed to reach equilibrium, and the gas phase concentration, $C_{g2}$, is measured. Derive an expression in terms of the above to determine the Henry's Constant, $H$, of the test chemical.

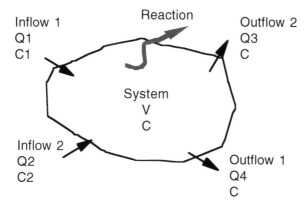

**Figure 4.4** Notations for a lumped system.

*Solution*

Let $M$ be the mass of chemical injected. This mass will distribute itself between the two phases to reach equilibrium conditions. MB equations can be written for the two bottles after they have reached equilibrium. Because there is no inflow or outflow and assuming no reactions within the bottles, the MB equations are as follows:

Bottle 1: $M = V_{l,1}C_{l,1} + V_{g,1}C_{g,1}$

Bottle 2: $M = V_{l,2}C_{l,2} + V_{g,2}C_{g,2}$

Because the bottles are at equilibrium, Henry's Law can be used to relate the phase concentrations in each bottle:

$$H = \frac{C_{g,1}}{C_{l,1}}$$

and

$$H = \frac{C_{g,2}}{C_{l,2}}$$

The liquid-phase concentrations $C_{l,1}$ and $C_{l,2}$ can now be eliminated between the four equations to yield:

$$H = \frac{\left(\dfrac{C_{g,1}}{C_{g,2}}\right)V_{l,1} - V_{l,2}}{V_{g,2} - \left(\dfrac{C_{g,1}}{C_{g,2}}\right)V_{g,1}}$$

**Worked Example 4.10**

A membrane separation system working on a batch mode consists of a well-mixed vessel containing a solution of a chemical at a concentration of $C_{r0}$, initially filled to a volume of $V_0$. The solution flows at a constant of $Q$ through the membrane. The concentration of the chemical in the permeate, $C_p$, is always related to the concentration in the vessel by a constant ratio of $p = C_p/C_r$. The volume remaining in the vessel at any time is $V_f$. Derive an expression for $C_p$ (see Figure 4.5).

*Solution*

Material balance for chemical: $\dfrac{d(V_rC_r)}{dt} = 0 - QC_p \pm 0$

From the information given: $C_p = pC_r$

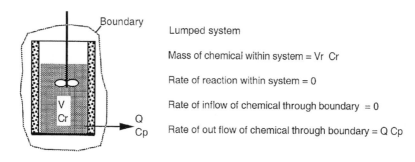

**Lumped system**

Mass of chemical within system = Vr Cr

Rate of reaction within system = 0

Rate of inflow of chemical through boundary = 0

Rate of out flow of chemical through boundary = Q Cp

**Figure 4.5** A membrane separation system.

One of the unknown concentrations, $C_r$, can be eliminated between the two equations to yield the following:

$$\frac{d\left(V_r \dfrac{C_p}{p}\right)}{dt} = -QC_p$$

Note that $V_r$ is a variable, and therefore, another relationship is needed to solve the above differential equation. The left-hand side of the above equation can be expanded using the chain rule:

$$\frac{d\left(V_r \dfrac{C_p}{p}\right)}{dt} = \frac{1}{p}\left[V_r \frac{dC_p}{dt} + C_p \frac{dV_r}{dt}\right]$$

Recognizing that the rate of decrease of volume inside the vessel, $(-dV_r/dt) = Q$, the flow rate out of the vessel, and the volume remaining at any time, $V_r = V_0 - Q_t$, the MB equation reduces to:

$$\frac{dC_p}{dt} = \frac{Q(1-p)}{V_0 - Qt} C_P$$

By separating the variables, the solution to the above can be developed as follows:

$$\frac{dC_p}{C_P} = \frac{Q(1-p)}{V_0 - Qt} dt$$

Integrating between the following initial conditions:

$$t = 0, \ C_{p0} = pC_{r0}$$

and

$$t = t, C_p = C_p$$

$$\int_{pCr0}^{Cp} \frac{dC_p}{C_p} = Q(1-p)\int_0^t \frac{1}{V_0 - Qt}\, dt$$

$$\ln\left(\frac{C_p}{pC_{r0}}\right) = (1-p)\ln\left(\frac{V_0 - Qt}{V_0}\right)$$

Hence, the final result is

$$C_p = pC_{r0}\left(\frac{V_0 - Qt}{V_0}\right)(1-p)$$

## EXERCISE PROBLEMS

4.1. A mixture of benzene and toluene exists as a gas-liquid system in equilibrium in a closed system at 80°C. The gas phase contains 65% mole benzene and 35% mole toluene. Assume that the vapor pressure of benzene = 756 and that of toluene = 287 mm Hg at 80°C.
   a. Find the total pressure of the system.
   b. What is the composition of the liquid phase?

4.2. In a gas-liquid binary system containing a solute A, the bulk mole fraction of A in the gas phase, $Y_1$, is 0.00035, and that in the liquid phase, $X_1$, is 0.03. Assume $X_1^* = 0.02$, $Y_i = 0.0004$, and $k_L = 0.01$.
   a. Determine the Henry's Constant of the solute.
   b. Construct the equilibrium diagram for the system.
   c. Determine $Y_1^*$ and $X_i$.
   d. Determine the overall mass transfer coefficients, $K_L$ and $K_G$.
   e. In what direction will the solute tend to flow?
   f. List four different equations to calculate the mass transfer flux and verify that all four equations yield the same result.

4.3. At an industrial facility, an open holding tank is used to equalize and store wastewaters for weekly discharge. Water samples taken from this tank indicate 2 mg dissolved oxygen in a 100-g water sample.
   a. Determine the direction of oxygen flux under this condition.
   b. When equilibrium is reached, what would be the dissolved oxygen concentration in the water (in mg/L)?
   c. To contain air emissions, this tank is now being covered, and the headspace is pressurized to 1.1 atm. Also, the headspace is enriched with oxygen. What should be the oxygen content (in %) in the headspace to ensure 20 mg/L of dissolved oxygen when the tank is discharged? Assume Henry's Constant of oxygen = $2.28 \times 10^7$ mm Hg/mole fraction.

4.4. A continuous flow of an aqueous waste stream containing toluene is pretreated by holding it in a closed tank and scavenging the headspace with a continuous flow of air.

a. Develop MB equations for the liquid and gas phases, and show that the concentration of toluene in the aqueous effluent, $C_{w,out}$, is given by the following expression:

$$C_{w,out} = C_{w,in}\left[1 + \cfrac{1}{\cfrac{Q_w}{V_w k} + \cfrac{Q_w}{K_{a-w}Q_a}}\right]^{-1}$$

where $C_{w,out}$ is the concentration of toluene in the influent, $Q_w$ is the flow rate of the waste stream, $Q_a$ is the flow rate of the air stream, $V_w$ is the volume of the waste in the holding tank, $k$ is the mass transfer coefficient, $K_{a-w}$ is the nondimensional concentration ratio form of the air-water partition coefficient of the VOC, $R$ is the universal gas constant, and $T$ is the temperature in the tank.

b. Hence, show that the concentration of the VOC in the off-gases from the tank, $C_{a,out}$, is given by:

$$C_{a,out} = C_{w,out}\left(\frac{Q_a}{kV_w} + \frac{1}{K_{a-w}}\right)^{-1}$$

4.5. Ammonia-nitrogen can exist in natural waters either in the ammonium ion form, $NH_4^+$, or in the un-ionized form, $NH_3$. The un-ionized form is of concern because of its toxicity to fish at concentrations around 0.01 mg N/L.

a. Derive a model to describe the amount of un-ionized ammonia as a % of total N, as a function of pH and temperature.

b. Hence, develop a nomograph to visualize the relationship between the above three variables.

4.6. Consider Worked Example 4.10.

a. How can the analysis be adapted if the chemical is being produced inside the vessel by a first-order reaction?

b. How can it be adapted if there is a constant inflow of the chemical, and the volume inside the vessel remains constant?

## APPENDIX 4.1 COMMON PARTITION COEFFICIENTS IN ENVIRONMENTAL SYSTEMS

| Phase I Content Measured as | Phase 2 Content Measured as | Partition Coefficient Symbol [Units] | Indication of |
|---|---|---|---|
| *Air* | *Water* | | |
| Partial pressure (atm.) | Molar conc. (moles/L) | $K_{a-w}$ (atm.-L/mole) | Volatility, strippability, solubility |
| Partial pressure (atm.) | Mole fraction (–) | (atm.) | |
| Mole fraction (–) | Mole fraction (–) | (–) | |
| Molar conc. (moles/L) | Molar conc. (moles/L) | (–) | |
| Mass conc. (mg/L) | Mass conc. (mg/L) | (–) | |
| *Air* | *Soil* | | |
| Partial pressure (atm.) | Mass conc. (mg/kg) | $K_{a-s}$ (atm.-kg/mg) | Emissions, adsorption, desorption |
| *Octanol* | *Water* | | |
| Molar conc. (moles/L) | Molar conc. (moles/L) | $K_{o-w}$ (–) | Bioaccumulation, hydrophobicity |
| *Soil* | *Water* | | |
| Mass conc. (mg/kg) | Mass conc. (mg/L) | $K_{s-w}$ (L/kg) | Adsorption, desorption, retardation |
| *Fish* | *Water* | | |
| Mass conc. (mg/kg) | Mass conc. (mg/L) | BCF (L/kg) | Bioaccumulation |

Partition Coefficient = $K_{1-2}$ = $\dfrac{\text{Phase 1 content}}{\text{Phase 2 content}}$

## CHAPTER 5

# Fundamentals of Engineered Environmental Systems

## CHAPTER PREVIEW

*Applications of the fundamentals of transport processes and reactions in developing material balance equations for engineered environmental systems are reviewed in this chapter. Alternate reactor configurations involving homogeneous and heterogeneous systems with solid, liquid, and gas phases are identified. Models to describe the performances of selected reactor configurations under nonflow, flow, steady, and unsteady conditions are developed. The objective here is to provide the background for the modeling examples to be presented in Chapter 8.*

## 5.1 INTRODUCTION

CHAPTER 4 contained a review of environmental processes and reactions. In this chapter, their application to engineered systems is reviewed. An engineered environmental system is defined here as a unit process, operation, or system that is designed, optimized, controlled, and operated to achieve transformation of materials to prevent, minimize, or remedy their undesired impacts on the environment.

The application and analysis of environmental processes and reactions in engineered systems follow the well-established practice of reaction engineering in the field of chemical engineering. While both chemical and environmental systems deal with processes and reactions involving liquids, solids, and gases, some important differences between the two systems have to be noted. Environmental systems are often more complex than chemical systems, and therefore, several simplifying assumptions have to be made in analyzing and modeling them. The exact composition and nature of the inflows are well defined in chemical systems, whereas in environmental systems, lumped surrogate measures are used (e.g., BOD, COD, coliform). The flow rates are often constant, steady, or predictable in chemical systems, whereas, in environmental systems, they are not, as a rule.

Engineered environmental systems are built up of *reactors*. A reactor is defined here as any device in which materials can undergo chemical, biochemical, biological, or physical processes resulting in chemical transformations, phase changes, or separations. The starting point in developing mathematical models of such reactors and systems is the material balance (MB). Principles of micro- and macro-transport theory and process/reaction kinetics (reviewed in Chapter 4) can be applied to derive expressions for inflows, outflows, and transformations to complete the MB equation. The mathematical form of the final MB equation can be algebraic or differential, depending on the nature of flows, reactions, and the type of reactor.

A complete analysis of reactors is beyond the scope of this book, and readers should refer to other specific texts on reactor engineering for further details. Excellent examples of such texts include those by Webber (1972), Treybal (1980), Levenspiel (1972), and Weber and DiGiano (1996).

## 5.2 CLASSIFICATIONS OF REACTORS

Reactors can be classified into several different types for the purposes of analysis and modeling. At the outset, they can be classified based on the type of *flow* and extent of *mixing* through the reactor. These factors determine the amount of time spent by the material inside the reactor, which, in turn, determines the extent of reaction undergone by the material. At one extreme condition, *complete mixing* of all elements within the reactor occurs; and at the other extreme, *no mixing* whatsoever occurs. The former type of reactors is referred to as *completely mixed reactors* and the latter, as *plug flow reactors*. Complete mixing here implies that *concentration gradients* do not exist within the reactor, and the reaction rate is the same everywhere inside the reactor. A corollary of this condition is that the concentration in the effluent of a completely mixed reactor is equal to that inside the reactor. In contrast, plug flow reactors are characterized by concentration gradients, therefore, they have spatially varying reaction rates within the reactor. Thus, completely mixed reactors fall under lumped systems, and plug flow reactors fall under distributed systems.

Reactors with either complete mixing on one extreme or no mixing on the other extreme are known as *ideal* reactors. Reactors in which some intermediate degree of mixing between the two extremes occurs are called *nonideal* reactors. While most reactors are analyzed and designed to be ideal, in practice, all reactors exhibit some degree of nonideality due to channeling, short-circuiting, stagnant regions, inlet/outlet effects, wall effects, etc. The degree of nonideality can be quantified through *residence time distribution* (RTD) studies. Even the best-designed reactors often exhibit some degree of non-ideality that requires complex models; hence, they are often approximated by modified ideal reactors. For example, large, nonideal continuous mixed-flow reactors (CMFRs) can be approximated by smaller, ideal CMFRs operating in series; large nonideal plug flow reactors (PFRs) may be approximated by ideal PFRs with dispersive transport added on. Thus, it is beneficial to fully appreciate ideal reactors and develop models for them so that large, full-scale reactors could be realistically designed, operated, and evaluated.

Ideal reactors can be further divided into homogeneous vs. heterogeneous, depending on the number of phases involved; flow vs. nonflow, depending on whether or not the flow of material occurs during the reaction; or steady vs. unsteady, depending on the time-dependency of the parameters. Illustrative applications of the fundamentals of environmental processes in homogeneous and heterogeneous reactors under flow, nonflow, steady, and unsteady conditions are presented in the following sections.

## 5.2.1 HOMOGENEOUS REACTORS

Homogeneous reactors entail reactions within one phase. Classification of some of the common homogeneous reactors is shown in Table 5.1.

The MB equation forms the basis for analyzing and modeling reactors. In the case of homogeneous reactors, bulk fluid flow characteristics and reaction kinetics at the macroscopic or reactor scale are primary factors to consider.

Table 5.1 Classification of Homogeneous Reactors

| Homogeneous Reactors | | | | |
|---|---|---|---|---|
| Nonflow Reactors | | | Flow Reactors | |
| Completely Mixed Batch Reactors (CMBR) | Sequencing Batch Reactors (SBR) | Completely Mixed Fed Batch Reactors (CMFBR) | Completely Mixed Flow Reactors (CMFR) | Plug Flow Reactors (PFR) |
| | | | With recycle | Without recycle |

## 5.2.2 HETEROGENEOUS REACTORS

Heterogeneous reactors entail reactions within two or more different phases such as gas-liquid, gas-solid, and liquid-solid systems. Classification of heterogeneous reactors commonly used in environmental studies is shown in Table 5.2.

While transport at the macro and reactor scales and reaction rates are the significant factors in homogeneous systems, micro- and macro-transport scales and inter- and intraphase mass transfer processes are significant in heterogeneous systems. As such, hydraulic retention times and reaction rate constants characterize homogeneous systems, and reaction rates and mass transfer coefficients characterize heterogeneous systems. The amounts of interfacial surface areas and path lengths for intraphase transport as well as bulk fluid dynamics contribute to the effectiveness of various heterogeneous reactor configurations.

## 5.3 MODELING OF HOMOGENEOUS REACTORS

In the following sections, development of the MB equation for various configurations of homogeneous reactors is summarized. The goal of this section is not to provide a formal treatment of reactor engineering, but instead to illustrate the different forms of MB equations, mathematical formulations, and the solution procedures that are involved in the modeling of common engineered environmental reactors.

Table 5.2 Classification of Heterogeneous Reactors

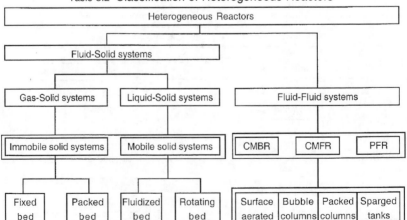

## 5.3.1 COMPLETELY MIXED BATCH REACTORS

In completely mixed batch reactors (CMBRs), the reactor is first charged with the reactants, and the products are discharged after completion of the reactions. During the reaction, inflow and outflow are zero, and the volume, $V$ ($L^3$), remains constant, but the concentration of the material undergoing the reaction changes with time, starting at an initial value of $C_0$. The MB equation for a CMBR during the reaction is as follows:

$$\frac{d(VC)}{dt} = -rV = -kCV \tag{5.1}$$

where $r$ is the rate of removal of the material by reactions ($ML^{-3}T^{-1}$), and $k$ is the first-order reaction rate constant ($T^{-1}$). The solution to the MB equation is as follows:

$$C = C_0 e^{-kt} \tag{5.2}$$

or

$$t = -\frac{1}{k} \ln\left(\frac{C}{C_0}\right) \tag{5.3}$$

where $C$ is the concentration of the material at any time, $t$, during the reaction.

## 5.3.2 SEQUENCING BATCH REACTOR

In sequencing batch reactors (SBRs), a sequence of processes can take place in the same reactor in a cyclic manner, typically starting with a fill phase. Reaction can occur during the fill phase of the cycle as the volume increases and can continue at constant volume after completion of the fill phase. On completion of the reaction, another process can take place, or the contents can be decanted to complete the cycle. The volume, $V_t$, at any time, $t$, during the fill phase $= V_0 + Qt$, where $V_0$ is the volume remaining in the reactor at the beginning of the fill phase (i.e., $t = 0$), and $Q$ is the volumetric fill rate ($L^3T^{-1}$). The MB equation during the fill phase, with reaction, for example, is as follows:

$$\frac{d(V_t C)}{dt} = rV_t + QC_{in} = -kCV_t + QC_{in} \tag{5.4}$$

which can be expanded to:

$$\frac{d}{dt}[(V_0 + Qt)C] = -kC(V_0 + Qt) + QC_{in} \tag{5.5}$$

The solution to the above MB equation is:

$$C = \frac{C_{in}}{(t + t_0)k} - \left[ \frac{C_{in}}{t_0 k} - C_0 \right] \frac{t_0}{(t + t_0)} e^{-kt} \tag{5.6}$$

where $t_0 = V_0/Q$ and $C_0$ is the concentration remaining in the reactor at $t = 0$. While the final result is difficult to interpret in the above form, a plot of $C$ vs. $t$ can provide more insight into the dynamics of the process. An Excel® model of the process is presented in Figure 5.1. A complete model for a biological SBR with Michaelis-Menten type reaction kinetics is detailed in Chapter 8, where the profiles of COD, dissolved oxygen, and biomass are developed employing three coupled differential equations.

## 5.3.3 COMPLETELY MIXED FLOW REACTORS

Completely mixed flow reactors (CMFRs) are completely mixed with continuous inflow and outflow. CMFRs are, by far, the most common environmental reactors and are often operated under steady state conditions, i.e., $d(\ )/dt = 0$. Under such conditions, the inflow should equal the outflow, while the active reactor volume, $V$, remains constant. A key characteristic of CMFRs is that the effluent concentration is the same as that inside the reactor. CMFRs can

| | | | |
|---|---|---|---|
| Influent concentration | Cin | 200 mg/L |
| Initial concentration | C0 | 5 mg/L |
| Vo/Q | to | 1 hr |
| First-order reaction rate constant k | | 0.7 1/hr |

`(Cin/((B10+to)*k))-((Cin/(to*k))-C0)*(to/(B10+to))*EXP(-k*B10`

| t | C |
|---|---|
| 0.0 | 5.00 |
| 0.5 | 58.60 |
| 1.0 | 73.16 |
| 1.5 | 74.99 |
| 2.0 | 72.16 |
| 2.5 | 67.70 |
| 3.0 | 62.83 |
| 3.5 | 58.11 |
| 4.0 | 53.73 |
| 4.5 | 49.76 |
| 5.0 | 46.21 |
| 5.5 | 43.04 |
| 6.0 | 40.21 |

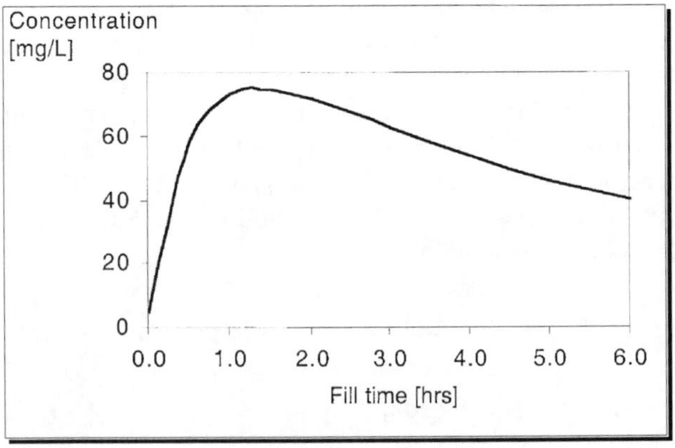

**Figure 5.1** Concentration profile in an SBR during the fill phase.

be characterized by their detention time, $\tau$, or the hydraulic residence time (HRT), which is given by $\tau = HRT = V/Q$. The material balance equation is as follows:

$$\frac{d(VC)}{dt} = rV + QC_{in} - QC \tag{5.7}$$

which at steady state reduces to:

$$0 = -kCV + QC_{in} - QC \tag{5.8}$$

whose solution is:

$$C = \frac{QC_{in}}{Q + kV} = \frac{C_{in}}{1 + k\left(\dfrac{V}{Q}\right)} = \frac{C_{in}}{1 + k\tau} \tag{5.9}$$

In some instances, multiple CMFRs are used in series, as shown in Figure 5.2, to represent a single nonideal reactor, or to improve overall performance, or to minimize total reactor volume.

For $n$ such identical CMFRs shown in Figure 5.2, the overall concentration ratio is related to the individual ratio of each reactor by the following series:

$$\frac{C_{out,n}}{C_{in}} = \left(\frac{C_1}{C_{in}}\right) \times \left(\frac{C_2}{C_1}\right) \times \ldots \left(\frac{C_{out,n}}{C_{n-1}}\right) \tag{5.10}$$

where $C_p$ is the effluent concentration of the $p$th reactor ($p = 1$ to $n$). Substituting from the result found above for a single CMFR into the above series gives the following:

$$\frac{C_{out,n}}{C_{in}} = \left(\frac{1}{1 + k\tau}\right)^n \quad \text{or, overall,} \quad \tau = n \left\{ \frac{\left[\dfrac{C_{in}}{C_{out,n}}\right]^{1/n} - 1}{k} \right\} \tag{5.11}$$

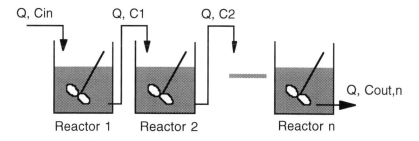

**Figure 5.2** CMFRs in series.

***Worked Example 5.1***

A wastewater treatment system for a rural community consists of two completely mixed lagoons in series, the first one of HRT = 10 days, and the second one of HRT = 5 days. It is desired to check whether this system can meet a newly introduced regulation of 99.9% reduction of fecal coliform by a first-order die off. The rate constant, $k$, for the die-off reaction has been found to be a function of HRT described by $k = 0.2\tau - 0.3$ (adapted from Weber and DiGiano, 1996).

*Solution*

Because Equation (5.11) assumes identical rate constants in all the reactors, it cannot be applied here. However, Equations (5.9) and (5.10) can be applied to yield the following:

$$\frac{C_{2,out}}{C_{in}} = \left(\frac{C_1}{C_{in}}\right)\left(\frac{C_{2,out}}{C_1}\right) = \left(\frac{1}{1 + k_1\tau_1}\right)\left(\frac{1}{1 + k_2\tau_2}\right)$$

Substituting the given data of: $\tau_1 = 10$ days, $\tau_2 = 5$ days, $k_1 = 0.2 * 10 - 0.3 = 1.7$, and $k_2 = 0.2 * 5 - 0.3 = 0.7$,

$$\frac{C_{2,out}}{C_{in}} = \left(\frac{1}{1 + (1.7)(10)}\right)\left(\frac{1}{1 + (0.7)(5)}\right) = 0.0123$$

and, hence, the percent reduction that can be achieved is 98.77%, which is less than the target of 99.9%.

One option for meeting the new standard is to construct a third lagoon in series. Its detention time can be determined as follows to achieve a reduction of 99.9%, or an overall concentration ratio of 0.001:

$$\frac{C_{3,out}}{C_{in}} = 0.001 = \left\{\left(\frac{C_1}{C_{in}}\right)\left(\frac{C_2}{C_1}\right)\right\}\left(\frac{C_{3,out}}{C_2}\right)$$

which gives

$$\left(\frac{C_{3,out}}{C_{in}}\right) = \frac{0.001}{\left\{\left(\frac{C_1}{C_{in}}\right)\left(\frac{C_2}{C_1}\right)\right\}} = \frac{0.001}{0.0123} = 0.081$$

Now, substituting this concentration ratio in Equation (5.9), and rearranging for $k\tau$,

$$k\tau = \left(\frac{C_2}{C_{3,out}}\right) - 1 = 12.35 - 1 = 11.35$$

and replacing $k$ in terms of the given function, results in a quadratic equation:

$$[0.2\tau - 0.3]\tau = 11.35$$

$$\text{or,} \quad 0.2\tau^2 - 0.3\tau = 11.35$$

giving a detention time of $\tau = 8.3$ days in the third lagoon to meet the new regulation.

## 5.3.4  PLUG FLOW REACTORS

In plug flow reactors (PFRs), elements of the material flow in a uniform manner, so that each plug of fluid moves through the reactor without intermixing with any other plug. As such, PFRs are also referred to as tubular reactors. The concentration within the reactor is, therefore, a function of the distance along the reactor. Hence, an integral form of the MB has to be used as shown in Figure 5.3 (see also Section 2.3 in Chapter 2).

For the element of length, $dx$, and area of cross-section, $A$, and velocity of flow, $u = Q/A$, the MB equation is:

$$\frac{d[(A dx)C]}{dt} = r(A dx) + QC - Q\left(C + \frac{dC}{dx} dx\right) \tag{5.12}$$

which at steady state yields:

$$0 = -(kC)(A dx) - Q\frac{dC}{dx} dx \tag{5.13}$$

or,

$$\frac{dC}{dx} = -k\left(\frac{1}{\dfrac{Q}{A}}\right)C = -\left(\frac{k}{u}\right)C \tag{5.14}$$

The solution to the above MB equation is as follows:

$$[\ln C]_{C_0}^{C_L} = -\left(\frac{k}{u}\right)\int_{x=0}^{x=L} dx = -\left(\frac{k}{u}\right)L \tag{5.15}$$

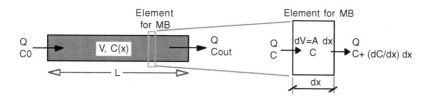

**Figure 5.3** Analysis of PFR.

or,

$$C_L = C_0 e^{-(k/u)L} = C_0 e^{-k\tau} \tag{5.16}$$

where $\tau = L/u$ is the hydraulic detention time, HRT.

### 5.3.5 REACTORS WITH RECYCLE

Reactors with some form of recycling often are advantageous over other reactor configurations, providing dilution of the feed and performance improvement. Recycling in CMFRs or PFRs is used more commonly in continuous flow heterogeneous reactors. Liquid recycling in CMFRs and PFRs, shown in Figure 5.4, can be modeled as follows by applying MB across the boundaries indicated:

#### 5.3.5.1 CMFR with Recycle

The MB equation for CMFR with recycle is as follows:

$$\frac{d(VC)}{dt} = QC_{in} + QRC - (Q + QR)C - rV = QC_{in} - (Q + kV)C \tag{5.17}$$

and, the solution to the MB equation at steady state is:

$$C = \frac{QC_{in}}{Q + kV} = \frac{C_{in}}{1 + k\left(\dfrac{V}{Q}\right)} = \frac{C_{in}}{1 + k\tau} \tag{5.18}$$

which is the same result as that found for CMFR without any recycle, Equation (5.9).

#### 5.3.5.2 PFR with Recycle

An integral MB equation has to be developed for PFR with recycle:

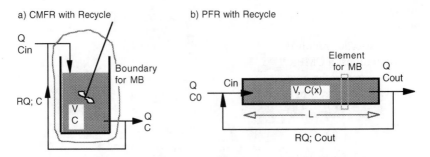

**Figure 5.4** CMFR and PFR with recycle.

$$\frac{d[(Adx)C]}{dt} = rAdx + (Q + RQ)C - (Q + RQ)\left(C + \frac{dC}{dx}dx\right) \quad (5.19)$$

which at steady state reduces to:

$$0 = -kAC - (Q + RQ)\frac{dC}{dx} \quad (5.20)$$

whose solution can be found as follows:

$$\int_{C_{in}}^{C_{out}} \frac{dC}{C} = \frac{-Ak}{Q(1 + R)} \int_0^L dx \quad (5.21)$$

$$\ln\left(\frac{C_{out}}{C_{in}}\right) = \frac{-Ak}{Q(1 + R)}L \quad (5.22)$$

or,

$$C_{out} = C_{in}e^{-[Ak/Q(1+R)]L} = C_{in}e^{-[k/u(1+R)]L} \quad (5.23)$$

A value for concentration $C_{in}$ at $x = 0$ can be found by applying an MB at the mixing point at the inlet:

$$C_{in} = \frac{QC_0 + RQC_{out}}{Q(1 + R)} \quad (5.24)$$

Hence, the final solution is as follows:

$$C_{out} = \left[\frac{e^{-[k/u(1+R)]L}}{1 + R - Re^{-[k/u(1+R)]L}}\right]C_0 \quad (5.25)$$

It can be noted that when $R = 0$, the above equation becomes identical to the one for the PFR without recycle, Equation (5.16).

### Worked Example 5.2

A first-order removal process is to be evaluated in the following reactor configurations: a CMFR, two CMFRs in series, three CMFRs in series, and a PFR. Compare the reactors on the basis of hydraulic retention time for removal efficiencies of 75, 80, 85, 90, and 95%.

### Solution

The HRT for a first-order process in a CMFR to achieve a removal efficiency of $\eta$ can be found by rearranging Equation (5.9) to get:

$$HRT = \left(\frac{1}{k}\right)\frac{\eta}{(1-\eta)}$$

The overall HRT for $n$ CMFRs in series can be found by rearranging Equation (5.11) to get:

$$HRT = n\left(\frac{1}{k}\right)\left\{\left[\frac{1}{(1-\eta)}\right]^{1/n} - 1\right\}$$

The HRT for a PFR can be found by rearranging Equation (5.16) to get:

$$HRT = \left(\frac{1}{k}\right)\ln\left[\frac{1}{(1-\eta)}\right]$$

Using the above equations, the following results can be obtained:

| | Overall HRT for | | | |
| Overall Efficiency | One CMFR | Two CMFRs | Three CMFRs | One PFR |
| --- | --- | --- | --- | --- |
| 75% | 30.0 | 20.0 | 17.6 | 13.9 |
| 80% | 40.0 | 24.7 | 21.3 | 16.1 |
| 85% | 56.7 | 31.6 | 26.5 | 19.0 |
| 90% | 90.0 | 43.2 | 34.6 | 23.0 |
| 95% | 190.0 | 69.4 | 51.4 | 30.0 |

## 5.4 MODELING OF HETEROGENEOUS REACTORS

The analysis and modeling of heterogeneous systems is often more complex than homogeneous systems. Furthermore, reactions in natural environmental systems are also typically heterogeneous. Hence, they are presented in this chapter, in somewhat more detail. However, because it is impossible to detail all the different reactor configurations, only a representative number of examples are presented.

### 5.4.1 FLUID-SOLID SYSTEMS

In liquid-solid reactors (and in gas-solid reactors), *contact* of liquids (or gases) with the *reactive sites* of the solid phase has to be facilitated by the transport processes. The reaction sites on the solid phase may be external at the surface and/or internal at the pores in the case of microporous or aggregated solids. The transport process and the reaction process occur in series, and therefore, the overall rate of gain or loss of material in the fluid phase will be *controlled* by transport alone or reaction alone or by both. The mass transfer process may be limited by *external resistance* due to boundary layers at the fluid-solid interface or by *internal pore resistances* in the case of microporous solids.

Examples of environmental liquid-solid reactors include adsorption, biofilms, catalytic transformations, and immobilized enzymatic reactions. Some of these reactors involve physical processes (e.g., carbon adsorption), while others involve chemical [e.g., UV-light-catalyzed reduction of Cr(VI) to Cr(III) by titanium dioxide] or biological reactions (e.g., removal of organics by biofilms). In this section, the development of two models of liquid-solid reactors is presented—one with biological reaction and one with a chemical reaction under nonideal conditions.

### 5.4.1.1 Slurry Reactor

Reactors used in activated sludge treatment, powdered activated carbon treatment (PACT), metal precipitation, and water softening can be categorized as slurry reactors. Here, biological flocs or precipitated solids represent the solid phase and act as catalysts to promote the reaction. These reactors are often modeled as CMFRs and are operated under steady state conditions. In this example, the activated sludge process is modeled, where the rate at which the dissolved substrate is consumed in the reactor is described using the Monod's expression. A typical CMFR-based activated sludge system is shown in Figure 5.5.

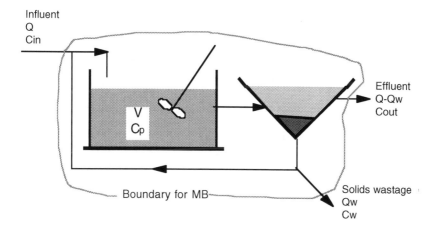

**Figure 5.5** Schematic of CMFR-based activated sludge process.

The MB equation for the above system under steady state conditions is as follows:

$$0 = QC_{in} - Q_wC_w - (Q - Q_w)C_{out} - raV \qquad (5.26)$$

where $r$ is the substrate uptake rate per unit reactive area $(ML^{-2}T^{-1})$, $a$ is the reactive area per unit volume of the reactor $(L^2L^{-3})$, and $V$ is the volume of the reactor $(L^3)$. The Monod's expression for reaction rate is:

$$r = r_{max}\left(\frac{C}{K_s + C}\right) \qquad (5.27)$$

where $r_{max}$ is the maximum surface substrate utilization rate $(ML^{-2}T^{-1})$ and $K_s$ is the half-saturation constant $(ML^{-3})$. The reactive surface area can be expressed as follows:

$$a = a_p\frac{C_p}{\rho_p} \qquad (5.28)$$

where $a_p$ is the reactive surface area per unit volume of the flocs $(L^2L^{-3})$, $C_p$ is the concentration of flocs in the reactor $(ML^{-3})$, and $\rho_p$ is the density of the flocs $(ML^{-3})$.

The following assumptions are made to simplify the analysis: the dissolved substrate does not undergo any reaction in the settling tank; concentration of the dissolved substrate, $C_w$, and the water flow rate, $Q_w$, in the solids wasting line are negligible when compared to the corresponding values in the influent and effluent; thus, $C = C_{out}$ and $Q - Q_w \sim Q$. Hence, combining the above Equations (5.26), (5.27), and (5.28) gives the following:

$$0 = QC_{in} - QC - r_{max}\left(\frac{C}{K_s + C}\right)\left(a_p\frac{C_p}{\rho_p}\right)V \qquad (5.29)$$

Even though the above expression is algebraic, it is somewhat difficult to solve for $C$ in the above form. If the reactor concentration, $C$, is small compared to $K_s$ (i.e., $K_s \gg C$), the reaction can be approximated by a first-order reaction with a rate constant $= (r_{max}/k_s)$, and Equation (5.29) can be readily solved to give the solution as follows:

$$C = C_{in}\left[1 + \left(\frac{r_{max}}{K_s}\right)a_pC_p\left(\frac{V}{Q}\right)\right]^{-1} = C_{in}\left[1 + \left(\frac{r_{max}}{K_s}\right)a_pC_{p\tau}\right]^{-1} \qquad (5.30)$$

## 5.4.1.2 Packed Bed Reactor

Packed bed reactors for contacting liquids (and gases) with solids are typically based on the PFR configuration. In essence, the fluid carrying the reactant(s) flows through a tubular reactor packed with the solid phase. The solid phase is retained within the reactor. The reaction occurs at the external or internal sites on the solid phase. Packed bed reactors can be engineered with immobile beds, expanded beds, or fluidized beds. In expanded bed reactors, the fluid velocity, $U$, is slightly greater than the settling velocity, $U_s$, of the solid particles, i.e., $U > U_s$; in fluidized bed reactors, the fluid velocity is significantly greater than the settling velocity, i.e., $U \gg U_s$.

The process model is essentially the same whether the bed is stationary, expanded, or fluidized. Reactor-scale macro-transport and element-scale micro-transport features of the three packed bed configurations are illustrated in Figure 5.6. In this example, the following assumptions are made: the reaction is first-order, and the fluid flow is nonideal, with advection and dispersion.

The steady state MB equation for the reactant in fluid phase can be developed as follows:

$$0 = QC - Q\left(C + \frac{dC}{dz}dz\right) + EA\frac{dC}{dz} - \left[EA\frac{dC}{dz} + \frac{d}{dz}\left(EA\frac{dC}{dz}\right)dz\right] - Na(A\,dz)$$

$$0 = Q\frac{dC}{dz}dz - \frac{d}{dz}\left(EA\frac{dC}{dz}\right)dz - Na(A\,dz)$$

$$0 = -u\frac{dC}{dz} - E\frac{d^2C}{dz^2} - Na \tag{5.31}$$

where $E$ is the dispersion coefficient ($L^2T^{-1}$), $a$ is the specific reactive area per unit volume of the reactor ($L^2L^{-3}$), $N$ is the interphase transport flux normal to

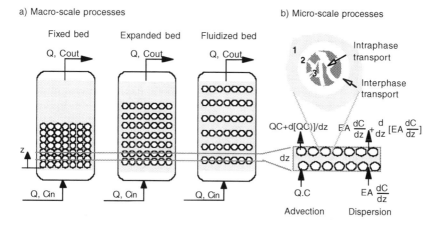

**Figure 5.6** Macro-scale and micro-scale representations of packed bed reactors (1—fluid phase; 2—laminar sublayer; 3—porous solid phase).

the particle surface $(ML^{-2}T^{-1})$, and $u = Q/A$ is the axial fluid velocity $(LT^{-1})$. The interphase flux can be expressed as follows:

$$N = k_f \Delta C = k_f(C - C_0) \tag{5.32}$$

where $k_f$ is the mass transfer coefficient $(LT^{-1})$ of the laminar sublayer and $\Delta C$ is the concentration difference $(ML^{-3})$ across the laminar sublayer around the particle $= (C - C_0)$, $C_0$ being the fluid phase concentration of the reactant immediately adjacent to the solid phase $(ML^{-3})$. The sublayer concentration difference can be expressed as follows:

$$\Delta C = (C - C_0) = \left( C - \left\{ \frac{k_f C}{k_f + k} \right\} \right) \tag{5.33}$$

where $k$ is the reaction rate constant, assuming it to be a first-order reaction. The above equations can now be combined, resulting in a second-order ODE. It has been solved between $z = 0$, $C = C_{in}$ and $z = L$, $C = C_{out}$, where $L$ is the length of the reactor. The final solution is as follows (Weber and DiGiano, 1996):

$$\frac{C_{out}}{C_{in}} = \frac{4\beta \, \exp\left( \dfrac{uL}{2E} \right)}{} \tag{5.34}$$

where

$$\beta = \sqrt{1 + 4k_f a\left( \frac{E}{uL} \right)}$$

## 5.4.2 FLUID-FLUID SYSTEMS

Environmental reactor configurations for processing gas-liquid systems include bubble columns (e.g., ozonation), packed towers (e.g., air-stripping), sparged tanks (e.g., activated sludge), and mechanical surface-aerated tanks (e.g., stabilization ponds). Some of these reactors involve only physical processes (e.g., volatilization in air-stripping towers), while some include chemical or biochemical reactions (e.g., ozonation, activated sludge). In this section, two examples illustrating the model development process for gas-liquid systems are detailed—one featuring a physical process and another featuring a biochemical process.

### 5.4.2.1 Packed Columns

Packed columns in which gas and liquid phases are contacted accompanied by transfers and/or reactions are common in many environmental and chemical engineering applications. In the environmental area, packed-column applications include stripping of volatile contaminants from water, oxygenation of wastewaters, adsorption of contaminants by activated carbon, stripping of nitrogen from wastewaters, and removal of organics in trickling filters. As an example, a model for a packed column used in air-stripping is presented next.

Packed columns for stripping volatile organic contaminants (VOCs), from groundwater, for instance, consist of countercurrent flow columns filled with inert packing media. Contaminated water is pumped to the top of the tower from where it flows under gravity through the packing media. Clean air is blown from the bottom of the column and flows upward. The contaminants are merely transferred from the aqueous phase to the gas phase without undergoing any chemical reaction. The countercurrent flow provides a large driving force for the transfer of the contaminant by volatilization, and the packed media provides a large interfacial area to enhance the transfer rate. Reactor-scale macro-transport and element-scale micro-transport representations of packed columns used in air-stripping are shown in Figure 5.7.

The packed column is often approximated as an ideal PFR, with the water and air streams flowing advectively with negligible mixing or dispersion. Hence, elemental MBs have to be written for the two phases. In this example, mole fractions are used rather than concentrations; the intention is not to cause any confusion, but to demonstrate that the fundamental concepts can be applied in any form. The steady state MB on the contaminant in the aqueous phase is as follows:

0 = Advective inflow – Advective outflow – Mass transferred to gas phase

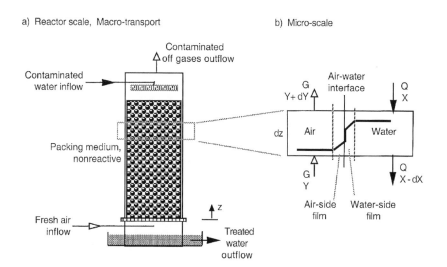

**Figure 5.7** Macro- and micro-transport processes in air-stripping.

$$0 = QX - Q(X - dX) - K_L(aA\,dz)(X - X^*) \tag{5.35}$$

where $Q$ is the molar flow rate $(MT^{-1})$; $X$ is the mole fraction of the contaminant in the aqueous phase, and $X^*$ is its mole fraction in the aqueous phase that would be in equilibrium with the mole fraction in the gas phase $(-)$; $K_L$ is the overall mass transfer coefficient with reference to the liquid-side film $(ML^{-2}T^{-1})$; $a$ is the interfacial area per unit volume of the reactor $(L^2L^{-3})$; and $A$ is the area of cross-section of the column $(L^2)$. Equation (5.35) can be simplified to determine the required height, $Z$, to achieve a certain removal of the contaminant:

$$z = \int_0^z dz = \left(\frac{Q}{K_L aA}\right)\int_{X_{in}}^{X_{out}} \frac{dX}{(X - X^*)} \tag{5.36}$$

The first term $Q/(K_L aA)$ in the right-hand side of the above equation has units of length (L) and is called the *height of a transfer unit* (HTU); the last integral term is a nondimensional quantity and is called the *number of transfer units* (NTUs). Hence, the column height can be expressed as

$$z = HTU \times NTU$$

To complete the integral, an expression for $X^*$ as a function of $X$ has to be established. Following the definition introduced in Chapter 4, $X^*$ and $Y$ are related through the air-water partition coefficient $K_{a-w}$ $(-)$ in the mole fraction ratio form.

$$X^* = \frac{Y}{K_{a-w}} \tag{5.37}$$

An equation relating $X$ and $Y$ can be developed by applying a macro-scale or overall MB across the reactor:

$$0 = QX_{in} + GY_{in} - QX_{out} - GY_{out} \tag{5.38}$$

Assuming that the airflow at the inlet is free of the contaminant (i.e., $Y_{in} = 0$), Equation (5.38) can be rearranged to get

$$Y_{out} = \frac{Q}{G}(X_{in} - X_{out}) \tag{5.39}$$

A similar MB between any point inside the reactor and the bottom of the reactor yields a similar result:

$$Y = \frac{Q}{G}(X - X_{out}) \tag{5.40}$$

Substituting for $Y$ in Equation (5.37) from Equation (5.40), and substituting the result into Equation (5.36), makes it finally amenable for integration:

$$z = \int_0^z dz = \left(\frac{Q}{K_L aA}\right)\int_{X_{out}}^{X_{in}} \frac{dX}{\left(X - \left(\frac{Q}{GK_{a-w}}\right)(X - X_{out})\right)} \qquad (5.41)$$

Hence,

$$z = \{HTU\}\{NTU\} = \left\{\frac{Q}{K_L aA}\right\}\left\{\left(\frac{R}{R-1}\right)\ln\left[\frac{X_{in}(R-1) + X_{out}}{RX_{out}}\right]\right\} \qquad (5.42a)$$

or, in terms of removal efficiency, $\eta$,

$$z = \left\{\frac{Q}{K_L aA}\right\}\left\{\left(\frac{R}{R-1}\right)\ln\left[\frac{R-\eta}{R(1-\eta)}\right]\right\} \qquad (5.42b)$$

where the nondimensional quantity, $R = (G/Q) K_{a-w}$, is known as the *stripping factor* (–). This final result is too complex to impart any intuitive feel for the process or its sensitivity to the different process variables. The relationship between tower height, removal efficiency, and HTU is illustrated in Figure 5.8 for $R = 15$. Similar graphical plots can be generated (with many of the software packages covered in this book) to gain insight into the process for optimal design and operation.

### 5.4.2.2 Sparged Tanks

Sparged tanks, in which gas and liquid phases are contacted accompanied by phase transfers and/or reactions are common in many environmental and

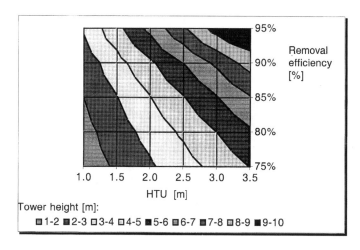

**Figure 5.8** Relationship between tower height, HTU, and removal efficiency.

chemical engineering applications. In the environmental area, sparged tank applications include oxygenation of wastewaters with air or high-purity oxygen, stripping of volatile contaminants from water, and removal of organics by ozonation. As an example, a model of a sparged tank for stripping VOCs is presented next.

In sparged tanks for aerating wastewaters, for example, oxygen is transferred from the gas phase into the liquid phase. At the same time, the gas phase can also strip dissolved gases or VOCs from the liquid phase. Such tanks are typically designed as continuous flow CMFRs and are operated under steady state conditions. The gas phase is introduced at the bottom of the tank, rising upward in a plug flow manner. In this section, modeling the stripping of VOCs in sparged tanks is outlined. A similar approach can be used to model oxygenation in aeration tanks (either with air or high-purity oxygen as the gas phase) and ozonation with gaseous ozone. Reactor-scale, macro-transport, and element-scale, micro-transport representations of sparged tanks used for stripping VOCs are as shown in Figure 5.9.

Consider a single rising bubble during its travel time $t_b$, and assume that its volume remains constant during its rise. The MB equation on VOCs inside the bubble is as follows:

$$\frac{d(V_b C_g)}{dt_b} = V_b \frac{dC_g}{dt_b} = K_G a V (C_g^* - C_g)$$

$$\frac{dC_g}{dt_b} = \frac{K_G a V}{V_b} (C_g^* - C_g) \tag{5.43}$$

where $V_b$ is the volume of the bubble at any instant ($L^3$); $K_G$ is the overall mass transfer coefficient relative to the gas phase ($LT^{-1}$); $a$ is the interfacial area per unit volume of reactor ($L^2 L^{-3}$); $C_g$ is the gas phase concentration of

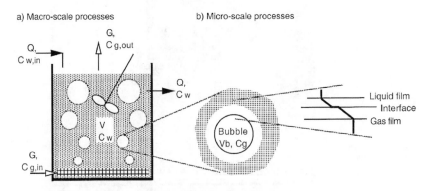

**Figure 5.9** Macro- and micro-transport processes in air-stripping in sparged tanks.

the VOC inside the bubble ($ML^{-3}$); and $C_g^*$ is the gas phase concentration of the VOC that would be in equilibrium with the liquid phase concentration of the VOC in the reactor, $C$. An expression for the bubble travel time and rise velocity can be derived as follows for solving the above MB equation:

$$t_b = \frac{z}{v_b} \rightarrow \frac{dC_g}{dt_b} = \frac{dC_g}{dz}\frac{dz}{dt_b} = \frac{dC_g}{dz}v_b \tag{5.44}$$

where $v_b$ is the terminal rise velocity of the bubble ($LT^{-1}$), and $z$ is the depth of the sparge tank (L). Hence, combining Equations (5.43) and (5.44),

$$\frac{dC_g}{dz} = \frac{K_G a V}{V_b v_b}(C_g^* - C_g) = \frac{K_G a V}{zG}(C_g^* - C_g) \tag{5.45}$$

where $G$ is the volumetric gas flow rate ($L^3 T^{-1}$). To eliminate $C_g^*$, the air-water partition coefficient, $K_{a-w}$ (Henry's Constant), can be used:

$$C_g^* = K_{a-w}C \tag{5.46}$$

where $C$ is the liquid-phase concentration of VOC in the reactor. The MB equation for VOCs in gas phase can now be solved as follows:

$$\frac{dC_g}{dz} = \left(\frac{K_G a V}{zG}\right)(K_{a-w}C - C_g)$$

$$\int_{C_{g,\text{in}}}^{C_{g,\text{out}}} \frac{dC_g}{(K_{a-w}C - C_g)} = \left(\frac{K_G a V}{zG}\right)\int_0^z dz$$

Because $C_{g,\text{in}} = 0$, the final result is:

$$C_{g,\text{out}} = K_{a-w}C\left[1 - \exp\left(-\frac{K_G a V}{G}\right)\right] \tag{5.47}$$

Finally, an overall steady state MB equation for the VOCs across the reactor can be written, assuming no other removal mechanism, and solved as follows:

$$0 = QC_{\text{in}} - QC - GC_{g,\text{out}}$$

$$0 = QC_{\text{in}} - QC - G\left\{K_{a-w}C\left[1 - \exp\left(\frac{K_G a V}{G}\right)\right]\right\} \tag{5.48}$$

to yield the concentration of VOCs that can be expected in the effluent of the CMFR:

$$C = \frac{C_{\text{in}}}{\left\{1 + \dfrac{GK_{a-w}}{Q}\left[1 - \exp\left(\dfrac{K_G a V}{G}\right)\right]\right\}} \tag{5.49}$$

This expression does not provide an intuitive appreciation of the process and the significance of the various parameters in the overall process. Performance curves generated in the following example can help in better understanding the process.

### Worked Example 5.3

An aeration tank with a volume of $3 \times 10^6$ ft$^3$ is receiving a flow of 5000 cfm. It is desired to evaluate the stripping efficiency of VOCs of a range of volatility in the aeration tank. Assuming a constant mass transfer coefficient for the bubble aeration device as 0.05 ft/min, develop a plot to show the relationship between the stripping efficiency, air-water partition coefficient, $K_{a-w}$, and the airflow-to-water flow ratio, $G/Q$.

### Solution

Equation (5.49) can be used to calculate the stripping efficiency and the contours. The stripping efficiency can be found as follows:

$$\eta = 100\left(\frac{C_{in} - C}{C_{in}}\right) = 100\left(1 - \frac{C}{C_{in}}\right)$$

$$= 100\left(1 - \left\{1 + \frac{GK_{a-w}}{Q}\left[1 - \exp\left(-\frac{K_G aV}{G}\right)\right]\right\}^{-1}\right)$$

| Overall mass transfer coefficient | KGa | 0.05 ft/min |
|---|---|---|
| Volume of reactor | V | 3E+06 cu ft |
| Wastewater flow rate | Q | 5000 cu ft/min |

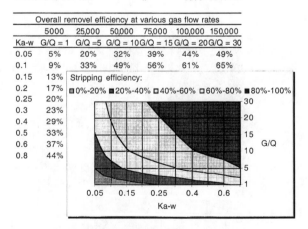

| Overall removel efficiency at various gas flow rates | | | | | |
|---|---|---|---|---|---|
| 5000 | 25,000 | 50,000 | 75,000 | 100,000 | 150,000 |
| Ka-w | G/Q = 1 | G/Q =5 | G/Q = 10 | G/Q = 15 | G/Q = 20 | G/Q = 30 |
| 0.05 | 5% | 20% | 32% | 39% | 44% | 49% |
| 0.1 | 9% | 33% | 49% | 56% | 61% | 65% |
| 0.15 | 13% | | | | | |
| 0.2 | 17% | | | | | |
| 0.25 | 20% | | | | | |
| 0.3 | 23% | | | | | |
| 0.4 | 29% | | | | | |
| 0.5 | 33% | | | | | |
| 0.6 | 37% | | | | | |
| 0.8 | 44% | | | | | |

Stripping efficiency:
□0%-20% ■20%-40% □40%-60% □60%-80% ■80%-100%

**Figure 5.10**

The above equation is implemented as a spreadsheet model as shown in Figure 5.10. This model calculates the stripping efficiency for $K_{a-w}$ values ranging from 0.05 to 0.8, at $G/Q$ values of 1, 5, 10, 15, 20, and 30. The contour plot generated from this model shows the required relationship between the three process parameters. Such a plot provides additional insight and aids in rapid evaluation of the overall process.

## EXERCISE PROBLEMS

5.1. Develop unsteady state MB equations for biomass and substrate concentrations, $X$ and $S$ (M/L$^{-3}$), respectively, in a batch bioreactor employing the following variables: maximum growth rate, $\mu_{max}$ (T$^{-1}$); half saturation constant, $K_s$ (ML$^{-3}$); first-order biomass death rate, $k_d$ (T$^{-1}$); first-order biomass respiration rate, $k_r$ (T$^{-1}$); and yield coefficient, $Y$ [M cells (M substrate)$^{-1}$]. Note that the biomass death process releases organic carbon back into the substrate pool, whereas the respiration process does not.

5.2. Using the same notation as in problem 5.1, develop MB equations for biomass and substrate in a CMFR with the following additional variables: the flow rate, $Q$ (L$^3$T$^{-1}$); influent substrate concentration, $C_{in}$ (M/L$^{-3}$); and volume of the reactor, $V$ (L$^3$). Assume negligible biomass concentration in the influent.

5.3. The aeration tank in a wastewater treatment plant is based on a CMFR design, using pure oxygen with bubble diffusers. A chemical plant wishes to discharge a waste stream containing 140 mg/L of toluene into the sewer system. The permit granted to the wastewater treatment plant (WWTP) limits the maximum concentration of toluene in its effluent to 1 mg/L. You are required to estimate the allowable discharge rate from the chemical plant, assuming that the WWTP influent previously carried no traces of toluene.

| | |
|---|---|
| Oxygen transfer efficiency of aerators | = 20% |
| Waste flow rate through aeration basin | = 1 MGD |
| Surface area of aeration basin | = 4000 sq ft |
| Hydraulic residence time | = 12 hrs |
| Oxygen requirement | = 200 mg/L of reactor volume |
| Henry's Constant of oxygen in wastewater | = 32 – |
| Henry's Constant of toluene in wastewater | = 0.45 – |
| DO level maintained in aeration basin | = 3.0 mg/L |
| $K_L$ for toluene/$K_L$ for oxygen | = 0.65 – |
| Average temperature | = 30°C |

If the WWTP used air instead of pure oxygen, do you think the chemical plant would be able to discharge at a higher rate than the one calculated above? Explain without any calculations.

5.4. Refer to Equation (5.35) where the overall mass transfer coefficient with reference to the liquid-side film, $K_L$, is dimensioned as ($ML^{-2}T^{-1}$), while in Equation (4.28), it is defined and dimensioned as ($LT^{-1}$). Reconcile these two forms of $K_L$.

5.5. Sparged tanks have been proposed for the removal of synthetic organic chemicals (SOCs) from water. Here, SOCs can be removed by two mechanisms—volatilization and oxidation. The oxidation process can be approximated by a first-order process of rate constant $k_{O_3}$. Following the approach and the notation used in developing Equation (5.49), and assuming the liquid phase to be completely mixed, develop a model to describe the effluent concentration of the SOC from the tank. The model should be in terms of the parameters $Q$, $G$, $V$, $K_{a-w}$, and $K_G a$ defined in Equation (5.49); and $\tau$, the hydraulic detention time; and $C_{O_3}$, the dissolved concentration of ozone in the reactor.

5.6. Continuing the above problem 5.4, construct MB equations for ozone in the gas and liquid phases. The rate of loss of ozone in the gas phase should equal the rate of consumption by the reaction in the liquid phase plus the rate of outflow in the effluent. Hence, derive an expression for $C_{O_3}$ that can be used in the result derived in the above problem 5.4.

## CHAPTER 6

# Fundamentals of Natural Environmental Systems

## CHAPTER PREVIEW

*This chapter outlines fluid flow and material balance equations for modeling the fate and transport of contaminants in unsaturated and saturated soils, lakes, rivers, and groundwater, and presents solutions for selected special cases. The objective is to provide the background for the modeling examples to be presented in Chapter 9.*

## 6.1 INTRODUCTION

IN this book, the terrestrial compartments of the natural environment are covered; namely, lakes, rivers, estuaries, groundwater, and soils. As in Chapter 4 on engineered environmental systems, the objective in this chapter also is to provide a review of the fundamentals and relevant equations for simulating some of the more common phenomena in these systems. Readers are referred to several textbooks that detail the mechanisms and processes in natural environmental systems and their modeling and analysis: Thomann and Mueller, 1987; Nemerow, 1991; James, 1993; Schnoor, 1996; Clark, 1996; Thibodeaux, 1996; Chapra, 1997; Webber and DiGiano, 1996; Logan, 1999; Bedient et al., 1999; Charbeneau, 2000; Fetter, 1999, to mention just a few.

Modeling of natural environmental systems had lagged behind the modeling of engineered systems. While engineered systems are well defined in space and time; better understood; and easier to monitor, control, and evaluate, the complexities and uncertainties of natural systems have rendered their modeling a difficult task. However, increasing concerns about human health and degradation of the natural environment by anthropogenic activities and regulatory pressures have driven modeling efforts toward natural systems. Better understanding of the science of the environment, experience from engineered systems, and the availability of desktop computing power have also contributed to significant inroads into modeling of natural environmental systems.

Modeling studies that began with BOD and dissolved oxygen analyses in rivers in the 1920s have grown to include nutrients to toxicants, lakes to groundwater, sediments to unsaturated zones, waste load allocations to risk analysis, single chemicals to multiphase flows, and local to global scales. Today, environmental models are used to evaluate the impact of past practices, analyze present conditions to define suitable remediation or management approaches, and forecast future fate and transport of contaminants in the environment.

Modeling of the natural environment is based on the material balance concept discussed in Section 4.8 in Chapter 4. Obviously, a prerequisite for performing a material balance is an understanding of the various processes and reactions that the substance might undergo in the natural environment and an ability to quantify them. Fundamentals of processes and reactions applicable to natural environmental systems and methods to quantify them have been summarized in Chapter 4. Their application in developing modeling frameworks for soil and aquatic systems is summarized in the following sections. Under soil systems, saturated and unsaturated zones and groundwater are discussed; under aquatic systems, lakes, rivers, and estuaries are included.

## 6.2 FUNDAMENTALS OF MODELING SOIL SYSTEMS

The soil compartment of the natural environment consists primarily of the unsaturated zone (also referred to as the vadose zone), the capillary zone, and the saturated zone. The characteristics of these zones and the processes and reactions that occur in these zones differ somewhat. Thus, the analysis and modeling of the fate and transport of contaminants in these zones warrant differing approaches. Some of the natural and engineered phenomena that impact or involve the soil medium are air emissions from landfills, land spills, and land applications of waste materials; leachates from landfills, waste tailings, land spills, and land applications of waste materials (e.g., septic tanks); leakages from underground storage tanks; runoff; atmospheric deposition; etc.

To simulate these phenomena, it is desirable to review, first, the fundamentals of the flow of water, air, and contaminants through the contaminated soil matrix. In the following sections, flow of water and air through the saturated and unsaturated zones of the soil media are reviewed, followed by their applications to some of the phenomena mentioned above.

## 6.2.1 FLOW OF WATER THROUGH THE SATURATED ZONE

The flow of water through the saturated zone, commonly referred to as groundwater flow, is a very well-studied area and is a prerequisite in simulating the fate, transport, remediation, and management of contaminants in groundwater. Fluid flow through a porous medium, as in groundwater flow, studied by Darcy in the 1850s, forms the basis of today's knowledge of groundwater modeling. His results, known as Darcy's Law, can be stated as follows:

$$u = \frac{Q}{A} = -K\frac{dh}{dx} \tag{6.1}$$

where $u$ is the average (or Darcy) velocity of groundwater flow ($LT^{-1}$), $Q$ is the volumetric groundwater flow rate ($L^3T^{-1}$), $A$ is the area normal to the direction of groundwater flow ($L^2$), $K$ is the hydraulic conductivity ($LT^{-1}$), $h$ is the hydraulic head ($L$), and $x$ is the distance along direction of flow ($L$). Sometimes, $u$ is referred to as *specific discharge* or *Darcy flux*. Note that the actual velocity, known as the *pore velocity* or *seepage velocity, $u_s$,* will be more than the average velocity, $u$, by a factor of three or more, due to the porosity $n$ (–). The two velocities are related through the following expression:

$$u_s = \frac{Q}{nA} = \frac{u}{n} \tag{6.2}$$

By applying a material balance on water across an elemental control volume in the saturated zone, the following general equation can be derived:

$$-\frac{\partial(\rho u)}{\partial x} - \frac{\partial(\rho v)}{\partial y} - \frac{\partial(\rho w)}{\partial z} = \frac{\partial(\rho n)}{\partial t} = n\frac{\partial(\rho)}{\partial t} + \rho\frac{\partial(n)}{\partial t} \tag{6.3}$$

where $u$, $v$, and $w$ are the velocity components ($LT^{-1}$) in the $x$, $y$, and $z$ directions and $\rho$ is the density of water ($ML^{-3}$). The three terms in the left-hand side of the above general equation represent the net advective flow across the element; the first term on the right-hand side represents the compressibility of the water, while the last term represents the compressibility of the soil matrix.

Substituting from Darcy's Law for the velocities, $u$, $v$, and $w$, under steady state flow conditions, the general equation simplifies to:

$$\frac{\partial}{\partial x}\left(K_x\frac{\partial h}{\partial x}\right) + \frac{\partial}{\partial Y}\left(K_y\frac{\partial h}{\partial y}\right) + \frac{\partial}{\partial z}\left(K_z\frac{\partial h}{\partial z}\right) = 0 \tag{6.4}$$

and by further simplification, assuming homogenous soil matrix with $K_x = K_y = K_z$, the above reduces to a simpler form, known as the Laplace equation:

$$\frac{\partial^2 h}{\partial x^2} + \frac{\partial^2 h}{\partial y^2} + \frac{\partial^2 h}{\partial z^2} = 0 \tag{6.5}$$

The solution to the above PDE gives the hydraulic head, $h = h(x, y, z)$, which then can be substituted into Darcy's equation, Equation (6.1) to get the Darcy velocities, $u$, $v$, and $w$.

### Worked Example 6.1

A one-dimensional unconfined aquifer has a uniform recharge of $W$ $(LT^{-1})$. Derive the governing equation for the groundwater flow in this aquifer. (The governing equation for this case is known as the Dupuit equation.)

*Solution*

The problem can be analyzed by applying a material balance (MB) on water across an element as shown in Figure 6.1. In this case, the water mass balance across an elemental section between 1-1 and 2-2 gives:

$$\text{Inflow at 1-1} + \text{Recharge} = \text{Outflow at 2-2}$$

$$(u \times h)|_1 + Wdx = (u \times h)|_2$$

Using Darcy's equation for $u$ and simplifying:

$$\left[\left(-K\frac{\partial h}{\partial x}\right) \times h\right]\bigg|_1 + Wdx = \left[\left(-K\frac{\partial h}{\partial x}\right) \times h\right]\bigg|_2$$

$$\left[\left(-K\frac{\partial h}{\partial x}\right) \times h\right]\bigg|_1 + Wdx = \left[\left(-K\frac{\partial h}{\partial x}\right) \times h\right]\bigg|_1 + \frac{\partial}{\partial x}\left[\left(-K\frac{\partial h}{\partial x}\right) \times h\right]dx$$

$$-\frac{\partial}{\partial x}\left[h\left(-K\frac{\partial h}{\partial x}\right)\right]dx + Wdx = 0$$

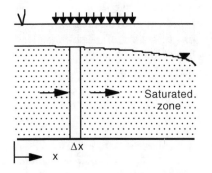
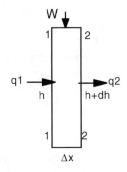

**Figure 6.1.** Application of a material balance on water across an element.

which has to be integrated with two BCs to solve for $h$. Typical BCs can be of the form: $h = h_o$ at $x = 0$; and, $h = h_1$ at $x = L$.

Following standard mathematical calculi, the above ODE can be solved to yield the variation of head $h$ with $x$. The result is a parabolic profile described by:

$$h^2 = h_o^2 + \frac{(h_L^2 - h_o^2)}{L} x + \frac{Wx}{K}(L - x)$$

The flux at any location can now be found by determining the derivative of $h$ from the above result and substituting into Darcy's equation to get:

$$u = \frac{K}{2L}(h_o^2 - h_L^2) + W\left(x - \frac{L}{2}\right)$$

### Worked Example 6.2

Two rivers, 1500 m apart, fully penetrate an aquifer with a hydraulic conductivity of 0.5 m/day. The water surface elevation in river 1 is 25 m, and that in river 2 is 23 m. The average rainfall is 15 cm/yr, and the average evaporation is 10 cm/yr. If a dairy is to be located between the rivers, which river is likely to receive more loading of nitrates that might infiltrate the soil.

### Solution

The equation derived in Worked Example 6.1 can be used here, measuring $x$ from river 1 to river 2:

$$u = \frac{K}{2L}(h_o^2 - h_L^2) + W\left(x - \frac{L}{2}\right)$$

The flow, $q$, into each river can be calculated with the following data:

$W$ = rainfall – evaporation = 15 – 10 = 5 cm/yr = $1.37 \times 10^{-4}$ m/day
$K = 0.5$ m/day, $L = 1500$ m, $h_o = 25$ m, $h_L = 23$ m

- river 1: $x = 0$

$$\therefore u = \frac{0.5 \frac{m}{day}}{2 * 1500\ m}(25\ m^2 - 23\ m^2) + \left(1.37 \times 10^{-4}\ \frac{m}{day}\right)\left(0 - \frac{1500}{2}\right)\ m$$

$$= -0.087\ \frac{m}{day}$$

The negative sign indicates that the flow is opposite to the positive $x$-direction, i.e., toward river 1.

- river 2:   $x = 1500$

$$\therefore u = \frac{0.5\frac{m}{day}}{2 * 1500 \text{ m}}(25 \text{ m}^2 - 23 \text{ m}^2) + \left(1.37 \times 10^{-4} \frac{m}{day}\right)\left(1500 - \frac{1500}{2}\right) \text{ m}$$

$$= 0.12 \text{ m/day}$$

Hence, river 2 is likely to receive a greater loading.

The problem is implemented in an Excel® spreadsheet to plot the head curve between the rivers. The divide can be found analytically by setting $q = 0$ and solving for $x$. The head will be a maximum at the divide. These conditions can be observed in the plot shown in Figure 6.2 as well, from which, at the divide, $x$ is about 650 m.

| | | |
|---|---|---|
| Rainfall | 15 | cm/yr |
| Evaporation | 10 | cm/yr |
| W | 5 | cm/yr |
| W | 1.37E-04 | m/day |
| K | 0.5 | m/day |
| L | 1500 | m |
| $h_o$ | 25 | m |
| $h_L$ | 23 | m |

**Flow calculations:**

River 1:
   $u = -0.0867$ m/day

River 2:
   $u = 0.1187$ m/day

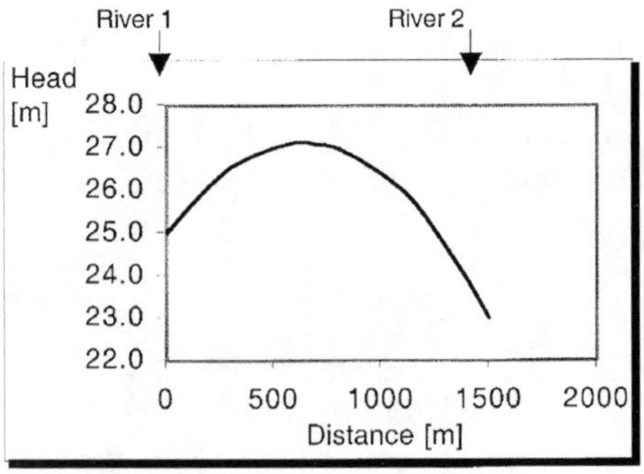

| x [m] | h [m] |
|---|---|
| 0 | 25.0 |
| 100 | 25.6 |
| 200 | 26.1 |
| 300 | 26.5 |
| 400 | 26.8 |
| 500 | 27.0 |
| 600 | 27.1 |
| 700 | 27.1 |
| 800 | 27.0 |
| 900 | 26.7 |
| 1000 | 26.4 |
| 1100 | 26.0 |
| 1200 | 25.4 |
| 1300 | 24.8 |
| 1400 | 24.0 |
| 1500 | 23.0 |

**Figure 6.2.**

## 6.2.2 GROUNDWATER FLOW NETS

The potential theory provides a mathematical basis for understanding and visualizing groundwater flow. A knowledge of groundwater flow can be valuable in preliminary analysis of fate and transport of contaminants, in screening alternate management and treatment of groundwater systems, and in their design. Under steady, incompressible flow, the theory can be readily applied to model various practical scenarios. Formal development of the potential flow theory can be found in standard textbooks on hydrodynamics. The basic equations to start from can be developed for two-dimensional flow as outlined below.

The continuity equation for two-dimensional flow can be developed by considering an element to yield

$$\frac{\partial u}{\partial x} + \frac{\partial v}{\partial y} = 0 \tag{6.6}$$

where $u$ and $v$ are the velocity components in the $x$- and $y$-directions. If a function $\psi(x,y)$ can be formulated such that

$$-\frac{\partial \psi}{\partial y} = u \quad \text{and} \quad \frac{\partial \psi}{\partial x} = v \tag{6.7}$$

then the function $\psi(x,y)$ can satisfy the above continuity equation. This function is called the *stream function*. This implies that if one can find the stream function describing a flow field, then the velocity components can be found directly by differentiating the stream function.

Likewise, another function $\phi(x,y)$ can be defined such that

$$-\frac{\partial \phi}{\partial x} = u \quad \text{and} \quad -\frac{\partial \phi}{\partial y} = v \tag{6.8}$$

which can satisfy the two-dimensional form of the Laplace equation for flow derived earlier, Equation (6.5). This function is called the *velocity potential function*. It can also be shown that $\phi(x,y) =$ constant and $\psi(x,y) =$ constant satisfy the continuity equation and the Laplace equation for flow. In addition, they are orthogonal to one another. In summary, the following useful relationships result:

in rectangular coordinates:

$$u = -\left(\frac{\partial \phi}{\partial x}\right) = \left(\frac{\partial \psi}{\partial y}\right) \quad \text{and} \quad v = -\left(\frac{\partial \phi}{\partial y}\right) = -\left(\frac{\partial \psi}{\partial x}\right)$$

in cylindrical coordinates:

$$u_r = -\left(\frac{\partial \phi}{\partial r}\right) = \frac{1}{r}\left(\frac{\partial \psi}{\partial \theta}\right) \quad \text{and} \quad u_\theta = -\frac{1}{r}\left(\frac{\partial \phi}{\partial \theta}\right) = -\left(\frac{\partial \psi}{\partial r}\right) \tag{6.9}$$

These functions are valuable tools in groundwater studies, because they can describe the path of a fluid particle, known as the streamline. Further, under steady flow conditions, the two functions, $\phi(x,y)$ and $\psi(x,y)$, are linear. Hence, by taking advantage of the principle of superposition, functions describing different simple flow situations can be added to derive potential and stream functions, and hence, the streamlines for the combined flow field.

The application of the stream and potential functions and the principle of superposition can best be illustrated by considering a practical example. The development of the flow field around a pumping well situated in a uniform flow field such as in a homogeneous aquifer is detailed in Worked Example 6.3, starting from the functions describing them individually.

***Worked Example 6.3***

Develop the stream function and the potential function to construct the flow network for a production well located in a uniform flow field. Use the resulting flow field to delineate the capture zone of the well.

*Solution*

Consider first, a uniform flow of velocity, $U$, at an angle, $\alpha$, with the $x$-direction. The velocity components in the $x$- and $y$-directions are as follows:

$$u = U \cos \alpha \quad \text{and} \quad v = U \sin \alpha$$

Substituting these velocity components into the above definitions for the potential and stream functions and integrating, the following expressions can be derived:

$$\phi = \phi_0 - U(x \cos \alpha + y \sin \alpha)$$

or

$$y = \frac{\phi_0 - \phi}{u \sin \alpha} - (\cos \alpha)x$$

$$\psi = \psi_0 + U(y \cos \alpha - x \sin \alpha)$$

or

$$y = \frac{\psi_0 - \psi}{u \cos \alpha} + (\tan \alpha)x$$

The results indicate that the stream lines are parallel, straight lines at an angle of $\alpha$ with the $x$-direction, which is as expected. In the special case where the flow is along the $x$-direction, for example, with $U = u$, the potential and stream functions simplify to:

$$\phi = \phi_0 - ux$$

and

$$\psi = \psi_0 - uy$$

Now, consider a well injecting or extracting a flow of $\pm Q$ located at the origin of the coordinate system. By continuity, it can be seen that the value of $Q = (2\pi\ r)\ u_r$, where $u_r$ is the radial flow velocity. Substituting this into the definitions of the potential and stream yields in cylindrical coordinates:

$$\phi = \phi_0 \pm \frac{Q}{2\pi}(\ln r)$$

and

$$\psi = \psi_0 \pm \frac{Q}{2\pi}(\theta)$$

which can be transformed to the more familiar Cartesian coordinate system as

$$\phi = \phi_0 \pm \frac{Q}{4\pi}[\ln (x^2 + y^2)]$$

and

$$\psi = \psi_0 \pm \frac{Q}{2\pi}\tan^{-1}\left(\frac{y}{x}\right)$$

The results confirm that the stream lines are a family of straight lines emanating radially from the well, and the potential lines are circles with the well at the center, as expected.

Because the stream functions and the potential functions are linear, by applying the principle of superposition, the stream lines for the combined flow field consisting of a production well in a uniform flow field can now be described by the following general expression:

$$\psi = \psi_0 + u(y \cos \alpha - x \sin \alpha) \pm \frac{Q}{2\pi}\tan^{-1}\left(\frac{y}{x}\right)$$

This result is difficult to comprehend in the above abstract form; however, a contour plot of the stream function can greatly aid in understanding the flow pattern. Here, the Mathematica® equation solver package is used to model this problem. Once the basic "syntax" of Mathematica® becomes familiar, a simple "code" can be written to readily generate the contours as shown in Figure 6.3. The capture zone of the well can be defined with the aid of this plot.

Notice that the *qTerm* is assigned a negative sign to indicate that it is pumping well. With the model shown, one can easily simulate various scenarios such as a uniform flow alone by setting the *qTerm* = 0 or an injection well alone by setting *u* = 0 and assigning a positive sign to the *qTerm* or by changing the flow directions through α.

### *Worked Example 6.4*

Using the following potential functions for a uniform flow, a doublet, and a source, construct the potential lines for the flow of a pond receiving recharge with water exiting the upstream boundary of the pond. Use the following values: uniform velocity of the aquifer, $U = 1$, radius of pond, $R = 200$, and recharge flow, $Q = 1000\pi$.

Uniform flow: $\phi = -ux$

Doublet: $\phi = \dfrac{R^2 x}{x^2 + y^2}$

Source: $\phi = -\dfrac{Q}{2\pi} \ln [\sqrt{x^2 + y^2}]$

```
In[7]:= u=1.0; qTerm = -500; α = Pi/6;
    ContourPlot[ψ[x,y] = u*(y*Cos[α] - x*Sin[α])+
    qTerm*ArcTan[y/x],
    {x,-2000,2000}, {y,-2000,2000},
    PlotPoints->100, Contours->20,ContourLines-
    >True,ContourShading->False, AspectRatio->0.6]
```

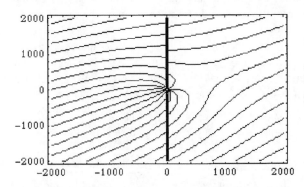

**Figure 6.3** Groundwater flow net generated by Mathematica®.

*Solution*

The potential function for the combined flow field is found by superposition:

$$\phi = -ux + \frac{R^2 x}{x^2 + y^2} - \frac{Q}{2\pi} \ln [\sqrt{x^2 + y^2}]$$

The contours of constant velocity potentials can be readily constructed with Mathematica® as shown in Figure 6.4. The plot shows that the stagnation point is upstream of the pond, implying that water is exiting the upstream boundary of the pond. This can be verified analytically by determining the velocity $u$ at $(-R,0)$ and checking if it is less than zero:

$$u_{(-R,0)} = -\left(\frac{\partial \phi}{\partial x}\right)_{(-R,0)} = \left\{ u\left[ 1 - \frac{R^2}{x^2 + y^2} + \frac{2R^2 x^2}{(x^2 + y^2)^2} \right] + \frac{Qx}{2\pi(x^2 + y^2)} \right\}_{(-R,0)}$$

The above reduces to:

$$u_{(-R,0)} = 2u - \frac{Q}{2\pi R}$$

```
In[76]:= u = 1; r = 200; q = 500;
        ContourPlot[
          φ[x, y] = -u x (1 - r^2/(x^2 + y^2)) - q*Log[{x^2 + y^2}^.5],
          {x, -1000, 1000}, {y, -1000, 1000}, PlotPoints → 50
          Contours → 30, ContourLines → True,
          ContourShading → False]
```

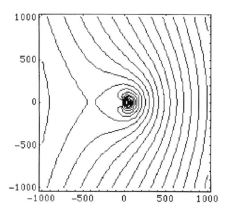

**Figure 6.4** Contours of potential function at $q = 300$.

giving the condition $Q > 4\pi Ru$ for flow to occur from the pond through its upstream boundary. In the above example, this condition is satisfied. The condition of $Q < 4\pi Ru$ can be readily evaluated by decreasing $Q$, for example, from $1000\pi$ to $600\pi$, and plotting the potential lines as shown in Figure 6.5.

The Mathematica® script can be easily adapted to superimpose the stream function on the velocity potential function as shown in Figure 6.6, for the two cases, to illustrate the orthogonality and to describe the flow pattern completely.

### 6.2.3 FLOW OF WATER AND CONTAMINANTS THROUGH THE SATURATED ZONE

Principles of groundwater flow and process fundamentals have to be integrated to model the fate and transport of contaminants in the saturated zone. Both advective and dispersive transport of the contaminant have to be included in the contaminant transport model, as well as the physical, biological, and chemical reactions.

As an initial step in the analysis of contaminant transport, the groundwater flow velocity components $u$, $v$, and $w$ $(LT^{-1})$, and the dispersion coefficients, $E_i$ $(L^2 T^{-1})$, in the direction $i$, can be assumed to be constant with

```
In[79]:= u = 1; r = 200; q = 300;
       ContourPlot[

       ϕ[x, y] = -u x (1 - r^2/(x² + y²)) - q * Log[(x^2 + y^2) ^ .5],

       {x, -1000, 1000}, {y, -1000, 1000}, PlotPoints → 50,
       Contours → 30, ContourLines → True,
       ContourShading → False]
```

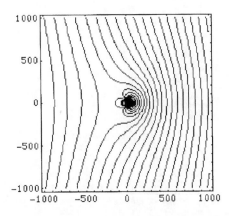

**Figure 6.5** Contours of potential function at $q = 300$.

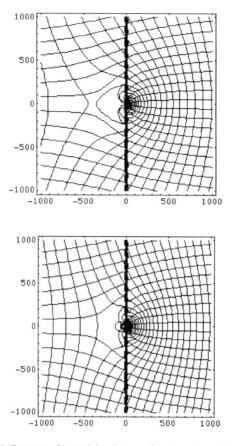

**Figure 6.6** Contours of potential and stream functions at $q = 500$ and $q = 300$.

space and time. A generalized three-dimensional (3-D) material balance equation can then be formulated for an element to yield:

$$\frac{\partial C}{\partial t} = -\left[ u\frac{\partial C}{\partial x} + v\frac{\partial C}{\partial y} + w\frac{\partial C}{\partial z} \right]$$

$$+ \left[ E_x\frac{\partial^2 C}{\partial x^2} + E_y\frac{\partial^2 C}{\partial y^2} + E_z\frac{\partial^2 C}{\partial z^2} \right] \pm \sum r \qquad (6.10)$$

where the left-hand side of the equation represents the rate of change in concentration of the contaminant within the element, the first three terms within the square brackets on the right-hand side represent the advective transport across the element in the three directions, the next three terms within the next square brackets represent dispersive transport across the element in the three directions, and the last term represents the sum of all physical, chemical, and biological processes acting on the contaminant within the element.

An important physical process that most organic chemicals and metals undergo in the subsurface is adsorption onto soil, resulting in their retardation relative to the groundwater flow. For low concentrations of contaminants, this phenomenon can be modeled assuming a linear adsorption coefficient and can be quantified by a *retardation* factor, $R$ (–), defined as follows:

$$R = 1 + \frac{K_d \rho_b}{n} \qquad (6.11)$$

where $K_d$ is the soil-water distribution coefficient $(L^3 M^{-1}) = S/C$, $n$ is the effective porosity (–), $\rho_b$ is the bulk density of soil $(ML^{-3}) = \rho_s(1 - n)$, $S$ is the sorbed concentration $(MM^{-1})$, and $\rho_s$ is the density of soil particles $(ML^{-3})$. Introducing the retardation factor, $R$, into Equation (6.10) and using retarded velocities, $u'$, $v'$, and $w'$ $(LT^{-1})$, and retarded dispersion coefficients, $E'_i$ $(L^2 T^{-1})$, results in the following:

$$\frac{\partial C}{\partial t} = -\left[ u'\frac{\partial C}{\partial x} + v'\frac{\partial C}{\partial y} + w'\frac{\partial C}{\partial z} \right]$$

$$+ \left[ E'_x\frac{\partial^2 C}{\partial x^2} + E'_y\frac{\partial^2 C}{\partial y^2} + E'z\frac{\partial^2 C}{\partial z^2} \right] \pm \frac{1}{R}\sum r \qquad (6.12)$$

The solution of the above PDE involves numerical procedures. However, by invoking simplifying assumptions, some special cases can be simulated using analytical solutions. These cases can be of benefit in preliminary evaluations, in screening alternative management or treatment options, in evaluating and determining model parameters in laboratory studies, and in gaining insights into the effects of the various parameters. They can also be used to evaluate the performance of numerical models. With that note, two simplified cases of one-dimensional flow (1-D) are given next.

*Impulse input, 1-D flow with first-order consumptive reaction*

A simple 1-D flow with first-order degradation reaction of the dissolved concentration, $C$, such as a biological process, with a rate constant, $k$ $(T^{-1})$, can be described adequately by the simplified form of Equation (6.12):

$$\frac{\partial C}{\partial t} = -\frac{u}{R}\frac{\partial C}{\partial x} + \frac{E}{R}\frac{\partial^2 C}{\partial x^2} - \frac{k}{R}C \qquad (6.13)$$

For example, an accidental spill of a biodegradable chemical into the aquifer can be simulated by treating the spill as an impulse load. For such an impulse input of mass $M$ (M), applied at $x = 0$ and $t = 0$ to an initially pristine aquifer over an area $A$ ($L^2$), the solution to Equation (6.13) has been reported to be as follows:

$$C = \frac{MR}{2A\sqrt{\pi E R t}} \left[ -\frac{\left(x - \frac{u}{R}t\right)^2}{\frac{4Et}{R}} \right] \left[ \exp\left(-\frac{k}{R}t\right) \right]$$ (6.14)

*Step input, 1-D flow with first-order consumptive reaction*

A continuous steady release of a biodegradable chemical originating at $t = 0$ into an initially pristine aquifer can also be described adequately by the simplified Equation (6.13). For example, leakage of a biodegradable chemical into the aquifer from an underground storage tank can be simulated by treating it as a step input load. The appropriate boundary and initial conditions for such a scenario can be specified as follows:

BC: $C(0, t) = C_o$ for $t > 0$ and $\frac{\partial C}{\partial x} = 0$ for $x = \infty$

IC: $C(x,0) = 0$ for $x \geq 0$

Under the above conditions, assuming first-order biodegradation of the dissolved concentration, $C$, the solution to Equation (6.13) has been reported to be as follows:

$$C = \frac{C_o}{2} \left\{ \exp\left[ \frac{(u - v)}{2D}x \right] \right\} \left\{ \text{erf}\left[ \frac{(Rx - vt)}{2\sqrt{DRt}} \right] \right\}$$

$$+ \frac{C_o}{2} \left\{ \exp\left[ \frac{(u - v)}{2D}x \right] \right\} \left\{ \text{erf}\left[ \frac{(Rx - vt)}{2\sqrt{DRt}} \right] \right\}$$ (6.15)

where $v = u \sqrt{1 + \frac{4kD}{u^2}}$

Additional numerical solutions for other special cases are included in Appendix 6.1 at the end of this chapter.

### Worked Example 6.5

An underground storage tank at a gasoline station has been found to be leaking. Considering benzene as a target component of gasoline, it is desired to estimate the time it would take for the concentration of benzene to rise to 10 mg/L at a well located 1000 m downstream of the station. Hydraulic conductivity of the aquifer is estimated as 2 m/day, effective porosity of the aquifer is 0.2, the hydraulic gradient is 5 cm/m, and the longitudinal dispersivity is 8 m. Assume the initial concentration to be 800 mg/L.

*Solution*

To be conservative, it may be assumed that degradation processes in the subsurface are negligible. This situation can be modeled using the equation given for Case 1 in Appendix 6.1:

$$C(L,t) = \frac{C_0}{2}\left\{\text{erfc}\left[\frac{L - ut}{2\sqrt{E_x t}}\right] - \exp\left[\frac{u}{E_x}L\right]\text{erfc}\left[\frac{L + ut}{2\sqrt{E_x t}}\right]\right\}$$

The second term in the above expression is normally smaller than the first term and, therefore, may be ignored. Using the given data,

$$u = K\frac{\left(\dfrac{dh}{dx}\right)}{n} = \left(2\ \frac{m}{day}\right)\frac{5\dfrac{cm}{m}\dfrac{m}{100\ cm}}{0.2} = 0.5\ \frac{m}{day}$$

$$E_x = \alpha_x u = 8\ m \times 0.5\ m/day = 4\ m^2/day$$

The above equation has to be solved for $t$, with $C(L,t) = 10$ mg/L; $C_o = 800$ mg/L; $L = 1000$ m; $u = 0.5$ m/day; and $E_x = 4$ m²/day:

$$100\ \frac{mg}{L} = \frac{800\ \dfrac{mg}{L}}{2}\left\{\text{erfc}\left[\frac{1000\ m - 0.5\ \dfrac{m}{day}t}{2\sqrt{4\dfrac{m^2}{day}t}}\right]\right\}$$

which gives

$$\text{erfc}\left[\frac{1000\ m - 0.5\ \dfrac{m}{day}t}{2\sqrt{4\dfrac{m^2}{day}t}}\right] = 0.25 \quad \text{or} \quad \frac{1000\ m - 0.5\ \dfrac{m}{day}t}{2\sqrt{4\dfrac{m^2}{day}t}} = 0.83$$

The above can be readily implemented in spreadsheet or equation solver-type software packages to solve for $t$. In this example, the Mathematica® equation solver package is used with the built-in *Solve* routine as shown:

In[16]:= **Solve[0.83 == (1000 - 0.5 \* t) / (2 \* √4 \* t).**

Out[16]= {{t → 1724.28}}

Hence, the time taken is 1724 days or 4.7 years.

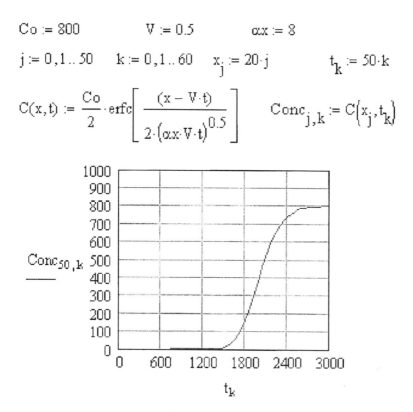

$$Co := 800 \qquad V := 0.5 \qquad \alpha x := 8$$

$$j := 0,1 .. 50 \quad k := 0,1 .. 60 \quad x_j := 20 \cdot j \qquad t_k := 50 \cdot k$$

$$C(x,t) := \frac{Co}{2} \cdot erfc\left[\frac{(x - V \cdot t)}{2 \cdot (\alpha x \cdot V \cdot t)^{0.5}}\right] \qquad Conc_{j,k} := C\left(x_j, t_k\right)$$

**Figure 6.7** Mathcad® model of benzene concentration.

This example is solved in another equation solver-type package, Mathcad®[6], as shown in Figure 6.7. Here, a plot of $C$ vs. $t$ is generated from the governing equation, from which it can be seen that the concentration at $x = 1000$ m will reach 100 mg/L after about 1750 days. The plot also shows that the peak concentration of 800 mg/L at the well will occur after about 3000 days.

## 6.2.4 FLOW OF WATER AND CONTAMINANTS THROUGH THE UNSATURATED ZONE

The analysis of water flowing past the unsaturated zone, infiltration for example, is an important consideration in irrigation, pollutant transport, waste treatment, flow into and out of landfills, etc. Analysis of flow through the unsaturated zone is more difficult than that of flow through the saturated

---

[6]Mathcad® is a registered trademark of MathSoft Engineering & Education Inc. All rights reserved.

zone, because the hydraulic conductivity, $K$, which is a function of moisture content, $\theta$ (–), changes as the water flows through the voids. Darcy's Law has been shown to be valid for vertical infiltration through the unsaturated zone, provided the $K$ is corrected for $\theta$ as follows:

$$w = -K_\theta \frac{\partial h}{\partial z} \tag{6.16}$$

where $h$ is the potential or head $= z + \omega$, $\omega$ being the tension or suction (L). For 1-D flow in the vertical direction, the water material balance for water simplifies to:

$$-\left[ \frac{\partial}{\partial z}(\rho w) \right] = \frac{\partial}{\partial t}(\rho\theta) \tag{6.17}$$

Assuming the soil matrix is nondeformable, the water is incompressible, and the resistance of airflow is negligible, the last two equations can be combined to yield the following result, known as the Richard's equation:

$$\frac{\partial\theta}{\partial t} = -\frac{\partial}{\partial z}\left[ K_\theta \frac{\partial\omega_\theta}{\partial z} \right] - \frac{\partial K_\theta}{\partial z} \tag{6.18}$$

This nonlinear PDE is very difficult to solve, except in simple special cases (Bedient et al., 1999). Analytical solutions reported for selected simplified cases are given next.

*Application to leachate concentration and travel time*
    The concentration of substances in leachates from surface spills and the time taken for the plume to break through the vadose zone can be estimated from the above analysis but with several simplifying assumptions as described by Bedient et al. (1994). First, the bulk concentration, $m$ $(ML^{-3})$, of the substance can be found in terms of the dissolved concentration, $C$ $(ML^{-3})$, taking into account its partitioning coefficients between air-water $(K_{a-w})$, pure organic phase-water $(K_{o-w})$, and soil-water $(K_{s-w})$, and the respective volumetric phase contents, $\theta_i$:

$$m = \Pi C = (\theta_w + K_{a-w}\theta_a + K_{o-w}\theta_o + K_{s-w}\rho_b)C \tag{6.19}$$

from which the total mass of the spill, $M$ (M), can be estimated from the area of contamination, $A$ $(L^2)$, and initial depth of the spill, $L_o$ (L):

$$M = AL_o(\Pi C) \tag{6.20}$$

Considering only the advective flow of the leachate from the contaminated zone at a Darcy velocity of $w$, a material balance on the substance yields:

$$\frac{dM}{dt} \rightarrow \frac{d}{dt}[AL_o\Pi C] \rightarrow AL_o\Pi\frac{dC}{dt} = -wAC \qquad (6.21)$$

which on integration with $C = C_o$ at $t = 0$ results in:

$$C = C_o \exp\left(-\frac{w}{L_o\Pi}t\right) \qquad (6.22)$$

The travel time can be estimated by assuming a power law model such as that of Brooks and Corey to relate unsaturated hydraulic conductivity, $K_\theta$, to the saturated conductivity, $K$. Assuming the Darcy velocity in the vertical direction to be the mean infiltration rate, $W$,

$$K_\theta = K\left(\frac{\theta - \theta_r}{n - \theta_r}\right)^\varepsilon \rightarrow \theta = \theta_r + (n - \theta_r)\left(\frac{W}{K}\right)^\varepsilon \qquad (6.23)$$

where $\theta$ is the volumetric water content, $\theta_r$ is the irreducible minimum water content, $n$ is the volumetric water content at saturation, and $\varepsilon$ is an experimental parameter. Now, using Darcy's Law and ignoring capillary pressures, the seepage velocity, $w_s$, can be approximated as follows:

$$w_s = \frac{W}{\theta_r + (n - \theta_r)\left(\dfrac{W}{K}\right)^{1/\varepsilon}} \qquad (6.24)$$

If $R$ is the retardation factor defined in Equation (6.11), the retarded velocity, $w'$ $(LT^{-1})$, can be obtained as

$$w' = \frac{W}{\theta_r + (n - \theta_r)\left(\dfrac{W}{K}\right)^{1/\varepsilon} + \rho_b k_b} \qquad (6.25)$$

and the travel time $t_{vz}$ through the vadose zone of length $L_{vz}$ (L) can be found from

$$t_{vz} = \frac{L_{vz}}{w'} = \frac{\left[\theta_r + (n - \theta_r)\left(\dfrac{W}{K}\right)^{1/\varepsilon} + \rho_b K_d\right]L_{vz}}{W} \qquad (6.26)$$

Also, if the substance undergoes a first-order degradation reaction at a rate of $k$ $(T^{-1})$ as it moves through the vadose zone, then the time course of the dissolved concentration as the plume reaches the groundwater table can be found from

$$C = C_o \exp\left\{-\left[\frac{w(t - t_{vz})}{L_o\Pi} + kt_{vz}\right]\right\} \qquad (6.27)$$

## 6.2.5 FLOW OF AIR AND CONTAMINANTS THROUGH THE UNSATURATED ZONE

The analysis of airflow through the unsaturated zone is of particular importance in remediation and treatment technologies, such as soil vapor extraction, air sparging, bioventing, and biofiltration; in quantifying emissions from land spills; etc. The general gas flow equation in this case can also be developed from Darcy's Law, and the solute transport equations can be developed from the material balance concept.

A simplified approach to model this phenomenon, specifically for application in soil vapor extraction, has been reported by Johnson et al. (1988). Assuming ideal gas characteristics and horizontal flow, the following equation has been derived for radial pressure distribution around a well in the unsaturated zone:

$$\frac{1}{r}\frac{\partial}{\partial r}\left(r\frac{\partial(\Delta P)}{\partial r}\right) = \left(\frac{\theta_a \mu}{\kappa P_{\text{atm}}}\right)\frac{\partial \Delta P}{\partial t} \tag{6.28}$$

where $r$ is the radial distance from the well (L); $\Delta P$ is the pressure deviation from atmospheric pressure, $P_{\text{atm}}$ ($\text{ML}^{-1}\text{T}^{-2}$); $\theta_a$ is the air-filled porosity (–); $\mu$ is the viscosity of air ($\text{MLT}^{-1}$); and $\kappa$ is the intrinsic permeability of soil ($\text{L}^2$). Equation (6.28) has been solved (Johnson et al., 1990) for the boundary condition:

$r$ = radius of the well, $R_w$       $P$ = well pressure, $P_w$

$r$ = radius of influence, $R_l$       $P$ = atmospheric pressure, $P_{\text{atm}}$

to get the following expression for the volumetric flow, $Q$, toward the well under pressure of $P_w$ at the well:

$$Q = \left(\frac{\pi H P_w \kappa}{\mu}\right)\left[\frac{1 - \left(\frac{P_{\text{atm}}}{P_w}\right)^2}{\ln\left(\frac{R_w}{R_l}\right)}\right] \tag{6.29}$$

The above result can be used to estimate the airflow $Q$, which in turn, can be used in estimating the transport of volatile contaminants during soil venting, for example. Johnson et al. (1988) proposed an equilibrium-based approach, where two scenarios may be considered:

- In the presence of the pure liquid phase of a contaminant in the contaminated zone, its concentration in the gas phase can be estimated using Raoult's Law to give the following:

$$C_a = \frac{(vp)(M_w)}{RT} \tag{6.30}$$

where *vp* is the vapor pressure of contaminant, $M_w$ is the molecular weight of contaminant, $R$ is the Ideal Gas Constant, and $T$ is the absolute temperature.

- In the case of absence of any liquid phase of the contaminant, its concentration in the gas phase can be estimated assuming partitioning of the contaminant between the soil, soil moisture, and gas phases:

$$C = \frac{K_{a-w}C_{\text{soil}}}{\left[\dfrac{K_{a-w}\theta_a}{\rho_{\text{soil}}} + \theta + k\right]} \tag{6.31}$$

where $K_{a-w}$ is the air-water partition coefficient, $C_{\text{soil}}$ is the concentration of the contaminant in the soil, $\theta$ is the moisture content, $k$ is the soil sorption constant, and $\rho_{\text{soil}}$ is the soil density.

The removal rate of contaminant from the soil, $M$, can now be estimated by multiplying the airflow, $Q$, from Equation (6.29), and the gas phase concentration, $C$, from Equation (6.30) or Equation (6.31), as $M = QC$.

## 6.3 FUNDAMENTALS OF MODELING AQUATIC SYSTEMS

Under aquatic systems, fundamental concepts relating to surface water bodies (lakes, rivers, and estuaries) are reviewed in this section.

### 6.3.1 LAKE SYSTEMS

In a simple analysis of the fate of a substance discharged into a lake, the lake can be characterized as a completely mixed system. While this assumption may be justified for shallow lakes, in more sophisticated analyses, larger lakes may be compartmentalized, and each compartment may be analyzed as completely mixed, with interactions between compartments. It is therefore useful to begin the modeling of lakes assuming completely mixed conditions. In that case, because the concentration of the substance in the outlet of the lake is the same as that inside the lake, the general form of the mass balance equation is as follows:

$$\frac{d(V_t C_t)}{dt} = Q_{\text{in},t}C_{\text{in},t} - Q_{\text{out},t}C_t \pm rV_t \tag{6.32}$$

where $V_t$ is the time-dependent volume of the lake ($L^3$), $C_t$ is the time-dependent concentration of the substance in the lake ($ML^{-3}$), $Q_{\text{in},t}$ is the time-dependent volumetric flow rate into the lake ($L^3T^{-1}$), $C_{\text{in},t}$ is the time-dependent concentration of the substance in the influent ($ML^{-3}$), $Q_{\text{out},t}$ is the time-dependent volumetric flow rate out of the lake ($L^3T^{-1}$), $r$ is the time-independent reaction rate of the substance in the lake ($ML^{-3}T^{-1}$), and $t$ is the time (T).

As a first step in modeling a lake and simulating its response to various perturbations, the above general equation may be simplified by invoking the following assumptions: the lake volume remains constant at $V$, the flow rate into and out of the lake are equal and remain constant at $Q$, the concentration of the substance in the influent remains constant at $C_0$, and all the reactions inside the lake are consumptive and of first order, of rate constant, $k$ ($T^{-1}$). With the above assumptions, using $C$ instead of $C_t$, Equation (6.32) reduces to:

$$V\frac{dC}{dt} = QC_{in} - QC - kVC \tag{6.33}$$

The above result can be rearranged to a standard mathematical form as follows:

$$\frac{dC}{dt} = \frac{QC_{in}}{V} - \frac{(QC + kVC)}{V} = \frac{QC_{in}}{V} - \left(\frac{Q}{V} + k\right)C$$

$$\frac{dC}{dt} + \left(\frac{Q}{V} + k\right)C = \frac{QC_{in}}{V}$$

$$\frac{dC}{dt} + \alpha C = \frac{W}{V} \tag{6.34}$$

where

$$\alpha = \left(\frac{Q}{V} + k\right) = \left(\frac{1}{\tau} + k\right) \tag{6.35}$$

$$\tau = \frac{V}{Q} \text{ is the hydraulic residence time (HRT) for the lake (T)} \tag{6.36}$$

$$W = QC_{in} \text{ is the load flowing into the lake } (MT^{-1}) \tag{6.37}$$

The simplifying assumptions make it easier to analyze the response of a lake under various loading conditions. Results from such simple analyses help in gaining a better understanding of the dynamics of the system and its sensitivity to the system parameters. With that understanding, further refinements can be incorporated into the simple model, if necessary. In the following sections, the application of simplified equations to several special cases mimicking real-life situations is illustrated.

### 6.3.1.1 Steady State Concentration

One of the elementary scenarios that can be simulated is the steady state condition to determine the in-lake concentration, $C_{ss}$, of a substance caused by a continuous, constant input load of $W$. The steady state condition is found by setting the first term in the left-hand side of Equation (6.22) to zero, yielding:

$$C_{ss} = \frac{W}{V\alpha} = \frac{W}{V\left(\frac{1}{\tau} + k\right)} = \frac{QC_{in}}{V\left(\frac{1}{\tau} + k\right)} = \frac{C_{in}}{(1 + k\tau)} \quad (6.38)$$

### 6.3.1.2 General Solution

A more general situation, where the application of a new step load to a lake with an initial concentration of $C_i$ causes a transient response, can be simulated by Equation (6.34), giving the solution as follows (Schnoor, 1996):

$$C = C_o \exp\left[-\left(\frac{1}{\tau} + k\right)t\right] + \frac{C_{in}}{1 + k\tau}\left\{1 - \exp\left[-\left(\frac{1}{\tau} + k\right)t\right]\right\} \quad (6.39)$$

It is apparent that, by setting $t$ to infinity, the first term in the right-hand side of Equation (6.39) dies off to zero, and the second term approaches the steady state value given by Equation (6.38).

The above equations can be applied to model the fate and transport of several water quality parameters such as pathogens, BOD, dissolved oxygen, nutrients, organic chemicals, metals, etc. Once the water column concentrations of these are established from the above equations, their impacts on other natural compartments of the lake systems such as suspended solids, biota, fish, sediments, etc., can also be analyzed. The above results can also be applied to simulate lakes in series, such as the Great Lakes, and compartmentalized lakes. Examples of such modeling are presented in the following chapters in this book.

### *Worked Example 6.6*

A lake of volume $V$ of $3.15 \times 10^9$ ft$^3$ is receiving a flow, $Q$, of 100 cfs. A fertilizer has been applied to the drainage basin of this lake, resulting in a load, $W$, of 1080 lbs/day to the lake. The first-order decay rate $K$ for this fertilizer is 0.23 yr$^{-1}$.

(1) Determine the steady state concentration of the chemical in the lake.
(2) If a ban is now applied on the application of the fertilizer, resulting in an exponential decline of its inflow to the lake, this decay can be assumed to be according to the equation $W = 1080e^{-\mu t}$, where $\mu = 0.05$ yr$^{-1}$ and $t$ is the time (yrs) after the ban. Develop a model to describe the concentration changes in the lake.

*Solution*

(1) The steady state concentration before the ban can be found from Equation (6.38). The detention time is first calculated as follows:

$$\tau = \frac{V}{Q} = \frac{3.15 \times 10^9 \text{ ft}^3}{100 \dfrac{\text{ft}^3}{\text{s}} \left( \dfrac{365 \times 24 \times 60 \times 60 \text{ s}}{\text{yr}} \right)} = 1 \text{ yr}$$

$$C_{ss} = \frac{W}{V\left(\dfrac{1}{t_d} + k\right)} = \frac{1080 \dfrac{\text{lbs}}{\text{day}} \left(\dfrac{454000 \text{ mg}}{\text{lbs}}\right) \left(\dfrac{365 \text{ day}}{\text{yr}}\right)}{3.15 \times 10^9 \text{ ft}^3 \left(\dfrac{28.32 \text{ L}}{\text{ft}^3}\right)\left(\dfrac{1}{1} + 0.23\right)\dfrac{1}{\text{yr}}} = 1.63 \dfrac{\text{mg}}{\text{L}}$$

(2) Assuming that the concentration in the lake has already reached 1.63 mg/L when the ban is introduced, a new MB equation has to be solved under the declining waste load. A modified form of Equation (6.34) can describe the system under the declining waste load:

$$\frac{dC}{dt} + \alpha C = \frac{W_o e^{-\mu t}}{V}$$

The above ODE has to be solved now, with an initial condition of $C = 1.63$ mg/L at $t = 0$. Following the standard mathematical calculi of using an integrating factor introduced in Section 3.3.2 in Chapter 3, the solution to the above can be found by applying Equation (3.18) with:

$$P(x) \equiv \alpha$$

and

$$Q(x) \equiv \frac{W_o}{V} e^{-\mu t}$$

Thus, the integrating factor is:

$$e^{\int P(x)dx} = e^{\int \alpha dt} = e^{\alpha t}$$

Therefore, the solution to the ODE is as follows:

$$C = \frac{\int \left\{ e^{\alpha t}\left(\dfrac{W_o e^{-\mu t}}{V}\right)\right\} dt + b}{e^{\alpha t}} = \frac{\left(\dfrac{W_o}{V}\right)\int \{e^{\alpha t}e^{-\mu t}\} dt + b}{e^{\alpha t}}$$

$$= \frac{\left(\dfrac{W_o}{V}\right)\int \{e^{(\alpha - \mu)t}\} dt + b}{e^{\alpha t}}$$

or

$$C = e^{-\alpha t}\left[\left(\left\{\frac{W_o}{V}\right\}\frac{1}{\alpha - \mu}\right)e^{(\alpha - \mu)t} + b\right]$$

where

$\alpha = [(1/\tau) + k] = 1.23 \text{ yr}^{-1}$

$\mu = 0.05 \text{ yr}^{-1}$

$W_o = 1080 \text{ lbs/day}$; and $V = 3.5 \times 10^9 \text{ ft}^3$

Hence, $C = e^{-1.23t}[1.70e^{1.18t} + b]$.

The integration constant $b$ can be found by substituting the initial condition:

$$C = 1.63 \text{ mg/L} \quad \text{at} \quad t = 0,$$

Therefore, $b = -0.07$.

Hence, the concentration in the lake, $C$ (mg/L), after the introduction of the ban can be described by the following equation:

$$C = e^{-1.23t}[1.70e^{1.18t} - 0.07]$$

In Chapter 7, several variations of this problem will be modeled with different types of software packages.

## 6.3.2 RIVER SYSTEMS

In a simple analysis of the fate of a substance discharged into swiftly flowing rivers and streams, river systems can be characterized as plug flow systems with negligible dispersion. In a more realistic analysis, dispersion might have to be included as described in the next section. Under the assumption of ideal plug flow conditions, the water flows without any longitudinal mixing in the direction of flow but with instantaneous mixing in all directions normal to the flow. Thus, the concentration of substances in the water column can vary with distance along the direction of flow as well as with time, but it is uniform at any cross-section. Such a condition can be analyzed by setting up a differential mass balance, assuming constant flow rate and first-order reactions:

$$\frac{d(A_x dx C)}{dt} = Q_t C - (Q_t + dQ)(C + dC) \pm kA_x dx C \pm S_D A_x dx \quad (6.40)$$

where $A_x$ is the area of flow at a distance $X$ from the origin ($L^2$), $x$ is the distance measured from origin along direction of flow (L), and, $S_d$ is the strength of a distributed source or a sink ($ML^{-3}T^{-1}$).

As a first step in modeling a river and simulating its response to various inputs, the above general equation may be simplified by invoking the following assumptions: the flow rate in the river is independent of $x$ and $t$, remaining constant at $Q$; the river is prismatic with the area of flow remaining constant at $A$; and all the reactions are of first order and are consumptive with a rate constant, $k$ $(T^{-1})$. With the above assumptions, Equation (6.40) can be reduced to:

$$\frac{\partial C}{\partial t} = -\left(\frac{Q}{A}\right)\frac{\partial C}{\partial x} - kC \pm S_D = -u\frac{\partial C}{\partial x} - kC \pm S_D \qquad (6.41)$$

where $u = Q/A$, the flow velocity in the river $(LT^{-1})$.

The above equation can be solved to predict the spatial and temporal variations of $C$ in the river system. Again, several simplified cases may be simulated to better understand the dynamics of the system.

### 6.3.2.1 Steady State without Source or Sink

The steady state condition can be simulated by setting the left-hand side of Equation (6.41) to zero and, in the absence of source or sink, by setting $S_D = 0$. Equation (6.41) then simplifies to the following:

$$\frac{\partial C}{\partial t} = 0 = -u\frac{\partial C}{\partial x} - kC \quad \text{or} \quad \frac{\partial C}{\partial x} = -\left(\frac{k}{u}\right)C \qquad (6.42)$$

Integrating the above between $x = 0$ to $x$ and $C = C_o$ yields the spatial variation of $C$ along the river:

$$C = C_o \exp\left(-\frac{k}{u}x\right) \qquad (6.43)$$

### 6.3.2.2 Steady State with a Distributed Source

When an infinitely long distributed source is included, Equation (6.41) under steady state conditions takes the form:

$$\frac{\partial C}{\partial x} + \left(\frac{k}{u}\right)C = \pm S_D \qquad (6.44)$$

Its solution with the boundary condition of $C = C_o = 0$ at $x = 0$ is:

$$C = \frac{S_D}{k}\left[1 - \exp\left(-\frac{x}{u}k\right)\right] \qquad (6.45)$$

The above equations can be applied to model the fate and transport of pathogens, BOD, dissolved oxygen, nutrients, organic chemicals, metals, etc., in river systems. The application of river models in the classical "DO sag curve" studies has been covered in several texts and is not repeated here.

Once the water column concentrations are established from the above equations, their impacts on other natural compartments of the system, such as suspended solids, biota, fish, sediment, etc., can also be modeled. As an example, the governing MB equations for a comprehensive, 11-variable nutrient-plants-oxygen model suitable for finite segment modeling reported by Thomann and Mueller (1987) are reproduced here. In the following equations, the advective and diffusive transports are combined and represented by $J(C_i)$ for the state variable $i$, for a segment of length $\Delta x$, where

$$J(C_i) = \left\{ -Q\left(\frac{\partial C_i}{\partial x}\right) + EA\left(\frac{\partial^2 C_i}{\partial x^2}\right) \right\}\Delta x \qquad (6.46)$$

- MB equation for phytoplankton $(C_1)$ in segment of volume $V$:

$$V\frac{dC_1}{dt} = J(C_1) + V(G_P - D_P)C_1 - VGC_1C_2 + W_1 \qquad (6.47)$$

where $G_P$ and $D_P$ are the growth and death rates of phytoplankton, $G$ is the grazing rate, $C_2$ is the zooplankton concentration, and $W_1$ is the input rate.

- MB equation for zooplankton $(C_2)$ in segment of volume $V$:

$$V\frac{dC_2}{dt} = J(C_2) + V(GC_1 - D_z)C_2 \qquad (6.48)$$

where $D_z$ is the respiration rate of zooplankton.

- MB equation for organic nitrogen $(C_3)$ in a segment of volume $V$:

$$V\frac{dC_3}{dt} = J(C_3) + VK_{3,4}C_3 + Va_{2,3}D_zC_2 - v_{n3}AC_3 + W_3 \qquad (6.49)$$

where $K_{3,4}$ is the conversion rate of organic nitrogen to ammonia, $a_{2,3}$ is the conversion factor for dead phytoplankton to organic nitrogen, $v_{n3}$ is the settling velocity, $A$ is the surface area, and $W_3$ is the input rate of organic nitrogen.

- MB equation for ammonia nitrogen $(C_4)$ in a segment of volume $V$:

$$V\frac{dC_4}{dt} = J(C_4) - Va_{1,4}G_PC_1 - VK_NC_4 + W_4 \qquad (6.50)$$

where $a_{1,4}$ is the conversion factor for ammonia nitrogen to phyto-plankton, and $K_N$ is the rate for conversion of ammonia to nitrate + nitrite.

- MB equation for nitrite + nitrate ($C_5$) in a segment of volume $V$:

$$V\frac{dC_5}{dt} = J(C_5) - Va_{1,5}G_PC_1 - VK_NC_4 + W_5 \qquad (6.51)$$

- MB equation for organic phosphorous ($C_6$) in a segment of volume $V$:

$$V\frac{dC_6}{dt} = J(C_6) - VK_{6,7}C_6 - Va_{1,6}D_PC_1 - Va_{2,6}D_zC_2$$
$$- v_{n6}AC_6 + W_6 \qquad (6.52)$$

- MB equation for orthophosphate phosphorous ($C_7$) in a segment of volume $V$:

$$V\frac{dC_7}{dt} = J(C_7) - Va_{1,7}G_PC_1 + W_7 \qquad (6.53)$$

- MB equation for silica ($C_8$) in a segment of volume $V$:

$$V\frac{dC_8}{dt} = J(C_8) - Va_{1,8}G_PC_1 + W_8 \qquad (6.54)$$

- MB equation for organic carbon ($C_9$) in a segment of volume $V$:

$$V\frac{dC_9}{dt} = J(C_9) - VK_dC_9 + Va_{1,9}D_PC_1$$
$$+ Va_{2,9}D_zC_2 - v_{n9}AC_9 + W_9 \qquad (6.55)$$

- MB equation for dissolved oxygen ($C_{10}$) in a segment of volume $V$:

$$V\frac{dC_{10}}{dt} = J(C_{10}) - VK_a(C_{s,10} - C_s) - Va_{10,9}K_dC_9 - Va_{10,4}K_NC_4$$
$$+ Va_{10,1}(G_P - D_P)C_1 - S_BA \qquad (6.56)$$

- MB equation for chlorides ($C_{11}$) in a segment of volume $V$:

$$V\frac{dC_{11}}{dt} = J(C_{11}) + W_{11} \qquad (6.57)$$

The above simultaneous differential equations are nonlinear and are of the second order. As such, numerical procedures have to be used to solve them. The steady state solutions can be found using the software packages discussed later in this text.

## 6.3.3 ESTUARY SYSTEMS

The model for estuaries or rivers with significant dispersion can be derived by adding the dispersive transport terms to the differential mass balance equation for the nondispersive river developed in the previous section. This analysis has already been presented in Section 2.3 in Chapter 2 for steady state conditions.

## 6.3.4 SPECIAL CASES IN RIVERS AND ESTUARIES WITH DISPERSION

The following two special cases of loading are of interest in water quality studies. They can be used to model tracer studies in determining the characteristics of rivers and estuaries. They can also be used to predict in-stream concentrations caused by accidental spills into rivers or estuaries.

- in-stream concentration due to an instantaneous spill of a chemical is as follows:

$$C_{(x,t)} = \frac{M}{2A\sqrt{\pi Et}} \exp\left[-\frac{(x - ut)^2}{4Et} - kt\right] \qquad (6.58)$$

where $M$ is the mass of the spill, $A$ is the area of flow, $E$ is the dispersion coefficient, $u$ is the flow velocity, and $k$ is the first-order reaction rate constant.

- in-stream concentration due to a constant of a chemical for a period of time $\tau$ is as follows:

$$C_{(x,t)} = \frac{C_o}{2} \exp\left(-\frac{kx}{u}\right)\left[\text{erf}\left(\frac{x - u(t - \tau)(1 + \eta)}{\sqrt{4E(t - \tau)}}\right)\right.$$

$$\left. - \text{erf}\left(\frac{x - ut(1 + \eta)}{\sqrt{4Et}}\right)\right] \qquad (6.59)$$

where

$$\eta = \frac{kE}{u^2}$$

The above equations are rather difficult to understand or visualize in the above form. Using mathematical software packages, however, the abstract equations can be modeled in different ways to gain valuable insight. As an example, a three-dimensional plot of the in-stream concentration given by Equation (6.58), as a function of time and distance developed with Mathcad® is shown in Figure 6.8.

$Q := 3.9$        $U := 0.47$            $M := 1.21 \cdot lb$        $E := 54$        $K := 0$

$A := \dfrac{Q}{U} \cdot ft^2$    $j := 1, 2 .. 10$        $k := 1, 2 .. 50$        $x_j := 10 \cdot j$        $t_k := 10 \cdot k$

$$C(x,t) := \dfrac{M}{2 \cdot A \cdot \sqrt{\pi \cdot E \cdot t}} \cdot e^{\left[ \dfrac{-(x - U \cdot t)^2}{4 \cdot E \cdot t} - K \cdot t \right]}$$        $CC_{j,k} := C\left(x_j, t_k\right)$

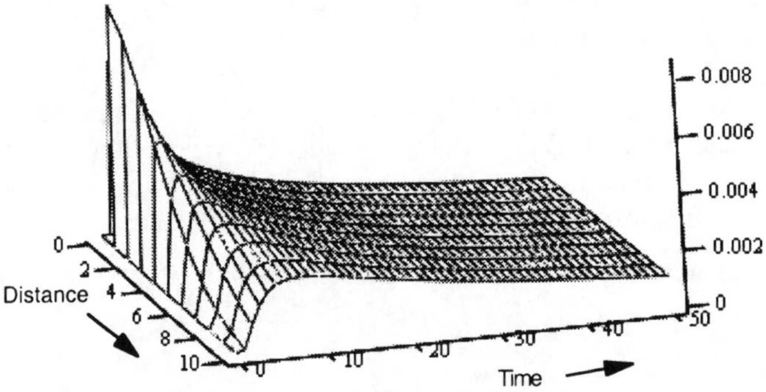

**Figure 6.8** Mathcad® model for concentration.

## EXERCISE PROBLEMS

6.1 Consider the Worked Example 6.4. Develop the stream function $\psi$, describing the problem and generating the stream lines.

6.2 A pump-and-treat system is being evaluated at a groundwater contamination site. The aquifer is unconfined with a uniform flow of 1 ft²/day. The pumping rate is 50 gpm (10,000 ft³/day). Estimate the maximum down gradient extent to which the drawdown cone of the well will capture the water. Note that the velocity potential for this problem can be constructed from the potential for a uniform flow and a source.

6.3 Agricultural drains are used to control groundwater level in crop cultivation as shown in Figure 6.9. Show that the height of the maximum saturated thickness between two drains is

$$h_m = \sqrt{h_d^2 + \frac{WL^2}{4K}}$$

where $W$ is the recharge and $K$ is the hydraulic conductivity of the soil.

6.4 Show that the following equation can be used to describe the seepage of water through an earthen dam:

$$U = \left(\frac{K}{2L}\right)(h_0^2 - h_L^2)$$

where $U$ is flow per unit width, $K$ is the hydraulic conductivity of the dam material, $L$ is its length, and $h_0$ and $h_L$ are the depth of water upstream and downstream of the dam.

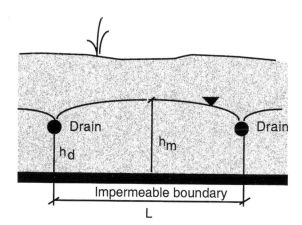

**Figure 6.9**

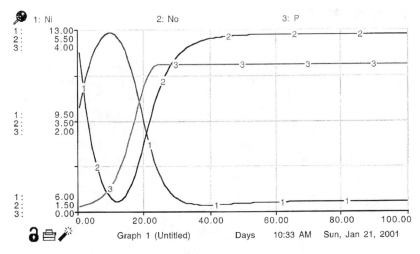

**Figure 6.10**

6.5   Consider the stream and potential functions for a source, a sink, and uniform flow. By placing the source and the sink symmetrically on the $x$-axis on either side of the $y$-axis, develop the following expressions for the stream and potential functions to describe the combined flow:

$$\psi = \frac{Q}{2\pi}(\theta - \theta_2) + Ur \sin \theta$$

$$\phi = \frac{Q}{2\pi} \ln\left(\frac{r_2}{r_1}\right) - Ur \cos \theta$$

Hence, plot the stream lines.

6.6   A simple model for phytoplankton and its interactions with nutrients in the epilimnion of Lake Ontario has been reported by Schnoor and O'Connor (1980). The model consisted of the following ODEs:

$$\frac{dP}{dt} = k_g N_i P - k_1 P - k_s P - \frac{P}{\tau}$$

$$\frac{dN_i}{dt} = \frac{W_1}{V} - k_g N_i P + k_o N_o - \frac{N_i}{\tau}$$

$$\frac{dN_o}{dt} = \frac{W_2}{V} + k_1 P - k_s N_o - k_o N_o - \frac{N_o}{\tau}$$

The above equations are simulated using the following data to generate the plots shown: $\tau = 534$ d, $k_1 = 0.25$ d$^{-1}$, $k_g = 0.0375$ d$^{-1}$, $k_5 = 0.024$ d$^{-1}$, $k_o = 0.14$ d$^{-1}$, $W_1/V = 0.0268$ µg/L-d, and $W_2/V = 0.0856$ µg/L-d.

Using one of the computational methods illustrated in Chapter 3, solve the above equations simultaneously to get the steady state concentrations, and compare them with the results from the plot in Figure 6.10.

## APPENDIX 6.1: ANALYTICAL SOLUTIONS FOR SPECIAL CASES OF GROUNDWATER CONTAMINATION

*Case 1:* Continuous Source of Conservative Tracer in One Dimension

For an infinite column of background concentration of zero and continuous input of a tracer at $x = 0$ at a concentration of $C_0$, the groundwater concentration at a point $x = L$ as a function of time is given by the following:

$$C(L,t) = \frac{C_o}{2}\left\{ \text{erfc}\left[\frac{L-ut}{2\sqrt{E_x t}}\right] - \exp\left[\frac{u}{E_x}L\right] \text{erfc}\left[\frac{L+ut}{2\sqrt{E_x t}}\right]\right\}$$

*Case 2:* Instantaneous Source of a Conservative Tracer in One Dimension

For an infinite column, of background concentration of zero, and slug input of a tracer of mass $M$, the concentration as a function of time and distance is given by the following:

$$C(x,t) = \frac{M}{2A\sqrt{\pi E t}}\left[-\frac{(x-ut)^2}{4Et}\right]$$

*Case 3:* Steady State Concentration in Two Dimension

Spatial distribution in two dimension for a plume that has stabilized is given by the following:

$$C(x,y) = \frac{C_0}{\sqrt{4\pi^2 E_x E_y}}\exp\left[\frac{ux}{2E_x}\right]K_o\left\{\frac{u^2}{4E_x}\left(\frac{x^2}{E_x^2} + \frac{y^2}{E_y^2}\right)\right\}$$

where $K_o$ is the modified Bessel function of second kind and zero order.

*Case 4:* Pulse Source in Two Dimension

If a tracer with a concentration of $C_0$ is injected over an area $A$ at $x = x_0$ and $y = y_0$ in a flow field, the concentration as a function of $x$, $y$, and $t$ is given by the following:

$$C(x,y) = \frac{C_o A}{\sqrt{16\pi^2 t^2 E_x E_y}} \exp\left[-\frac{[(x-x_0)-ut]^2}{4E_x t} - \frac{(y-y_0)^2}{4E_y t}\right]$$

*Case 5:* Pulse Radioactive Source in Three Dimension

If a radioactive tracer of first-order decay rate of $\lambda$ with a concentration of $C_0$ and volume $V_0$ is released into a flow field at $t = 0$, the downstream con- centration as a function of $x$, $y$, and $t$ is given by the following:

$$C(x,y,z,t) = \frac{C_0 V_0}{8(\pi t)^{3/2}\sqrt{E_x E_y E_z}} \exp\left[-\frac{(x-ut)^2}{4E_x t} - \frac{y^2}{4E_y t} - \frac{z^2}{4E_z t} - \lambda t\right]$$

*Case 6:* Plane Source of Reactive Substance in Three Dimension

$$C(x,y,z,t) \quad \frac{C_0}{8}\exp\left\{\frac{x}{2\alpha_x}\left[1-\sqrt{\left(1+\frac{4\lambda\alpha_x}{u}\right)}\right]\right\} \times$$

$$\text{erfc}\left\{\frac{x-ut\sqrt{\left(1+\frac{4\lambda\alpha_x}{u}\right)}}{2\sqrt{\alpha_x ut}}\right\}\left\{\text{erf}\left[\frac{\left(y+\frac{y}{2}\right)}{2\sqrt{\alpha_x}x}\right] - \text{erf}\left[\frac{\left(y-\frac{y}{2}\right)}{2\sqrt{\alpha_x}x}\right]\right\} \times$$

$$\left\{\text{erf}\left[\frac{(z+Z)}{2\sqrt{\alpha_z}x}\right] - \text{erf}\left[\frac{(z-Z)}{2\sqrt{\alpha_z}x}\right]\right\}$$

## CHAPTER 7

# Software for Developing Mathematical Models

## CHAPTER PREVIEW

*In this chapter, three distinct types of commercially available software packages for model development are identified: spreadsheet-based, equation solver-based, and dynamic simulation-based packages. Selected examples of software packages belonging to these three types are discussed. Some of their features, merits, and limitations are illustrated by applying them to the same common water quality modeling problem. The objective is to provide an overview of the available software and their capabilities so that readers can make their own choices appropriate to their modeling goals.*

## 7.1 INTRODUCTION

THE low-cost availability of high-performance desktop computer hardware and equally powerful software applications in recent years has fostered extensive use of computer-based simulation models in all fields of science and engineering. The benefits of computer-based simulation models in understanding, analyzing, and predicting the behavior of complex and large-scale natural and engineered systems in a safe, timely, and cost-effective manner have been well recognized.

A large number of professionally developed, special purpose modeling and simulation programs are available commercially and also as shareware/freeware. Most such programs in use today have been developed using traditional computer programming languages such as Fortran, Pascal, C, BASIC, etc.

End users often adapted these models in a black-box manner, feeding the required input parameters in a specific format and obtaining values for certain output parameters preset by the programmer. The source code of these models are normally not accessible to the users for them to modify the program, if necessary, to meet their particular needs.

Often, instructors, students, and professionals face situations when pre-developed packages may not be flexible, readily available, or adequate for special purpose applications. In such cases, it may be desirable for them to develop their own programs or "models" to meet their individual needs. Using traditional programming languages to develop simulation models demands considerable computer programming expertise in addition to the subject matter expertise. Even for subject matter experts with advanced programming skills, developing special purpose models using traditional programming languages may prove to be very tedious and time consuming. To justify the cost of model development by this approach, the models must have wide applicability and/or a large market.

Today's computer users, who are familiar with software driven by menu commands and/or mouse clicks with on-screen, format-free data entry via dialog boxes, expect similar features in simulation models as well. They also expect built-in features for graphical presentation and statistical analyses of the model outputs as well as for sharing the outputs among other software applications and other computers locally and globally. Using traditional programming languages to develop models that incorporate these features demands considerable programming effort.

Recognizing a need for model development tools for nonprogrammers, software developers have introduced a new breed of applications that nonprogrammers can quickly learn and use for developing their own simulation models. These applications can be thought of as software tool kits for building or "authoring" special purpose simulation models for limited uses and/or users. Such applications enable authors to create professional-quality simulation models cost effectively, at a fraction of the time, requiring minimal programming skills.

These applications are user-friendly in that model developers can build models of varying complexity using familiar operators and mathematical logic in contrast to using syntax and programming logic with traditional programming languages. They also feature several built-in routines for plotting, animation, statistical analysis, and presentation that the model developers can adapt and incorporate into their models with ease. Above all, these models can be configured to be flexible so that end users can, within certain limits, modify and adapt basic models developed by others to suit their own needs. Such flexibility allows models to be refined, fine-tuned, and upgraded as the user becomes more familiar with the problem being studied.

Currently available software packages suitable for authoring computer-based models can be categorized into three distinct types: spreadsheet-based applications, equation solver-based packages, and dynamic simulation-based packages. Examples of packages falling into the three categories that are selected for illustration in this book are summarized in Appendix 7.1. In this chapter, some of the salient features of these example packages are outlined and illustrated by applying them in modeling the same problem.

## 7.2 SPREADSHEET-BASED SOFTWARE

Spreadsheet-based software such as Excel®, Quattro® Pro[7], and Lotus®[8], have been available for such a long time that their features and capabilities are almost identical. Even though spreadsheets were originally designed as electronic accounting books for financial analysis, they have evolved into powerful mathematical tools and have been successfully adapted by modelers to simulate a wide range of scientific and engineering phenomena. In a way, spreadsheet applications are to numbers what word processors are to text.

A worksheet in a spreadsheet application takes a tabular format, consisting of columns, designated by alphabets, and rows, designated by numerals. The intersection of any column (for example, column P) with any row (for example, row 6) forms a cell, identified by its column heading and row number as P6 in this example. Users can click inside any cell and enter text, numeric constants or variables, built-in functions or logical expressions, or custom equations, all of which in turn can use or refer to constants, variables, or even functions and custom equations, contained in other cells. Custom equations can be embedded in the cells by typing them directly on the screen using standard mathematical notations; the terms in the equations can refer to cells that, in turn, contain the constants, variables, functions, or other custom equations. Links between the cells are "live" in that any change entered into a cell will instantly update the values of all the cells that depend on that cell as well as plots generated from those cells.

Spreadsheets feature a wide range of built-in mathematical, statistical, and logical functions that users can enter into cells using standard mathematical notations with minimal syntax. They also contain built-in, menu-driven routines for storing, formatting, and sorting data; plotting graphs; performing trials and solutions; data analysis and curve fitting; exporting/importing data, etc. In addition, they also include an English-like scripting language that advanced users can adapt to write special purpose functions called "macros"

---

[7]Quattro® Pro is a registered trademark of Corel Corporation. All rights reserved.
[8]Lotus® is a registered trademark of Lotus Development Corporation. All rights reserved.

to further enhance the capabilities of spreadsheets. A unique feature of Excel[®] is that it has a comprehensive set of drawing tools built in, for authors to use to create sophisticated graphic objects from within the program.

In building a spreadsheet model, the author places model parameters and the model equations or *formulas* into the cells. The cells that carry the formulas receive inputs from other cells that are linked to them, and they perform the operation specified. The numerical results calculated at any cell can, in turn, be utilized by other linked cells to perform further calculations or plotting. Thus, a spreadsheet model contains essentially a series of cells carrying the model input parameters, the governing equations, and the model outputs. The cells display only the numerical values or logical expressions generated by the embedded equations and not the equations themselves. However, the embedded equation in any cell will be displayed in the formula bar when the user clicks the mouse in that cell. Alternatively, all the cells in a model can be set to display only their respective formulas all at once through the menu by opening *Tools* > *Preferences* > *View* and checking the *Formulas* button under the *Window Options*.

Spreadsheet applications such as Excel[®] are relatively inexpensive, commonly available, easy to learn and use, and very fast and powerful for algebraic operations. Applications of Excel[®] in modeling environmental systems have been well documented (Hardisty et al., 1993; Gottfried, 2000). However, the program is limited by its inability to maintain internal dimensional homogeneity; incapability of making symbolic manipulations or calculus-based operations such as integration and differentiation, and lack of advanced math functions such as complex numbers, gamma functions, numerical procedures, etc. With advanced spreadsheet programming skills, however, some of these limitations may be circumvented.

## 7.3 EQUATION SOLVER-BASED SOFTWARE

Several types of equation solving packages with powerful mathematical capabilities have become available in the past two decades. Some of the more common ones are Mathcad[®], Mathematica[®], MATLAB[®], and TK Solver. They are to mathematical equations what spreadsheets are to numbers and word processors are to text.

### 7.3.1 Mathcad[®]

Mathcad[®] is designed to process equations numerically and symbolically, especially for performing engineering analysis. In setting up simulation models with Mathcad[®], the model constants and variables are declared first at the top of a worksheet followed by the governing equations. The governing equations are entered exactly as they would appear in a textbook, with the unknown on the left-hand side of the equation and known variables on the right-hand side. Within a worksheet, Mathcad® performs the calculations from top to bottom, handling equations numerically or symbolically, but only in one direction. All the variables in the right-hand side of the equation must have been previously "declared" above the equation and should have known values. Mathcad® has a built-in capability for calculus and matrix operations, complex numbers, series calculations, advanced vector graphics, animations, curve fitting, interpolating, and numerical procedures. It does not feature any built-in drawing tools (like those in Excel®), but it allows users to import graphics from other applications.

In contrast to the other common equation solving packages, Mathcad® features several unique attributes. Mathematical notations and equations appear on the worksheet in true symbolic form and are live—any changes made to the constants or variables above an equation are immediately reflected in the results of that equation as well as in the results of all the equations and plots that depend on them. It does not require any syntax or code to build up models involving elementary algebra or basic calculus. More importantly, it maintains dimensional homogeneity, which is of particular benefit in several engineering applications when common parameters are quantified in mixed units. Use of Mathcad® in developing models for solving engineering problems has been documented in several reports, papers, and books (e.g., Pritchard, 1999).

## 7.3.2. Mathematica®

Mathematica® is structured so that the kernel that performs all the computations is separate from the front end where the user interacts. The front end can be set for text-based interface or graphic interface. With the text-based interface, the users interact primarily through the keyboard; with the graphic interface, users interact through palettes, buttons, menus, etc. This graphic interface supports a high degree of interactivity and is available for the PC and Macintosh platforms. Users' inputs and the program's outputs, graphics, and animation can be integrated in the notebook to generate publication-quality materials.

Users interact at the notebook level by entering equations or expressions; the front end passes the information to the kernel where the computations are completed, then receives the results from the kernel and presents them at the notebook level. User inputs can be in the form of regular keyboard characters or in the two-dimensional form with special characters selected from a palette. The latter form is known as the standard form. Outputs at the notebook level are always expressed in the standard form. Whenever users input an equation or expression, they are labeled as *In[#]*, and the corresponding outputs are labeled as *Out[#]*.

The above interface features with the *basic input palette* are illustrated in Figure 7.1. When the user inputs line *In[1]* and line *In[3]* using the regular keyboard characters, Mathematica® echoes those as outputs in line *Out[1]* and in line *Out[3]*, respectively, in the standard from. The two-dimensional expression entered in line *In[6]* in the standard form using the basic input palette is echoed in line *Out[6]* in the standard form. The two-dimensional expression entered in line *In[11]* is *evaluated* and returned as an output in the standard form.

The last two lines in Figure 7.1 illustrate just one of the many powerful features of Mathematica® in performing mathematical operations in symbolic form. The rich collection of features in the program includes numerous built-in analytical and numerical functions and procedures, plotting, animation, and visualization tools, expendability, etc. It also has the ability to rearrange and backsolve an equation for one variable at a time.

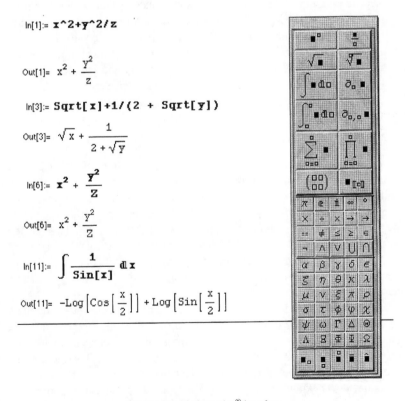

Figure 7.1 Mathematica® interface.

## 7.3.3 MATLAB®

MATLAB® is yet another equation processing application. It is based on the matrix approach (the name is derived from MATrix LABoratory) and integrates computation, programming, and visualization to model complex systems mathematically and graphically. Internally, all variables in the MATLAB® environment are treated as matrices, with $1 \times 1$ matrices considered scalers and one-column or one-row matrices considered vectors. The matrix approach enables complex calculations to be implemented efficiently and compactly in an elegant manner.

While MATLAB®'s capabilities and features are similar to those of Mathematica®, its interface is quite different. Modelers and users interact with MATLAB® through the *Command* window. Any valid expression entered in the command window is interpreted and evaluated. The expressions can consist of operators, functions, and variables. The evaluation results in the answer in a matrix form. The expressions can be entered line-by-line in the *Command* window for immediate evaluation.

When building models in MATLAB®, a sequence of commands is assembled to input the model parameters and to translate the mathematical model into a MATLAB® "script." When interacting at the *Command* window, this script has to be typed in every time the model is run with different inputs. This can be tedious when the script contains a large number of commands and inputs. To avoid this problem, the program allows modelers to store the script in a specific file format called the M-File and call that file by name from the *Command* window to be executed with different inputs. At run time, the call to the script in the M-File can pass inputs and receive the results generated by the script through *Arguments* to the call.

The scripts can be written using the built-in MATLAB® language or traditional programming languages. By combining built-in functions and custom-defined functions, almost any type of problem can be modeled with the MATLAB® system. The functionality of the M-Files can be twofold. One type of M-Files, called *Script* files, can perform a desired operation without returning any result. For example, a sequence of valid expressions that generate a plot fall into this type. The second type of M-Files, called *Function* files, can receive some variables as inputs, perform some calculations, and return the result to the Command window or to other M-Files for further processing. A sequence of expressions that receives a set of numbers to calculate their mean and standard deviation and return the results is an example of the second type.

This interface might sometimes be confusing to users not familiar with the MATLAB® environment. To minimize this confusion, modelers can make use of the program's built-in tools to build graphical user interfaces (GUIs) so that users can run MATLAB® models interactively, without having to know the program's environment. MATLAB®, however, lacks the graphical two-dimensional interface for entering equations (as in Mathcad® or Mathematica®) and the ability to maintain dimensional homogeneity (as in Mathcad®). Several references to the use of MATLAB® in engineering model building can be found in the literature, and books on the use of MATLAB® are also plentiful (e.g., Palm III, 2001).

### 7.3.4  TK Solver

The interface in TK Solver is built of sheets. The governing equations are first entered in algebraic form in the *Rule Sheet*. The equations can be entered in any sequence. TK Solver will automatically create a *Variables Sheet*, listing all the variables contained in the equations in the *Rule Sheet*. The *Variables Sheet* consists mainly of an *Input* column, a *Name* column, an *Output* column, and a *Unit* column. The known values for the variables are entered by the user under the *Input* column of the *Variables Sheet*, and TK Solver will solve all of the equations and return the unknown variables in the *Output* column of the *Variables Sheet*.

The model is constructed by assembling the algebraic equations, just as in the Excel® spreadsheet package. Unlike Mathcad®, TK Solver does not have the built-in ability to perform the calculations in consistent units. However, the model developer can specify the units in which each variable is displayed in the *Variable* sheet (*Display* unit) and the units in which it is used in the equations (*Calculation* units). To implement consistency, the developer has to fill in another sheet, called the *Unit* sheet, where specific conversion factors between the *display* units and *calculation* units are specified.

Variables in TK Solver can take single values or a *List* of multiple values. When all the variables take single values, the model can be *Solved* performing one set of calculations to return single values for the unknowns. When one or more variables take multiple values, as in a list, the model can be *ListSolved* to perform multiple calculations to return multiple values for the unknowns. The *List* feature is used to solve equations for a series of input values for one variable at a time and generate a series of output values of the other variables to plot graphs, for example.

While TK Solver can also be categorized among the equation solving packages, it has the unique capability of inversion or backsolving the same basic model. For example, if an equation has $n$ variables, of which any of the $(n - 1)$ are known, TK Solver can solve the equation for the $n$th unknown using just one statement of the equation. Traditional computer programs and math packages would need $n$ assignment statements to do the same calculations. If a problem involves $m$ independent equations with an average of $n$ variable each, then a total of $m * n$ number of assignment statements would be required in traditional programs and math packages, while TK Solver would solve the problem with just *m* equations. When the equations are inter-related, building a general model to simulate the system by traditional methods becomes complex.

## 7.4 DYNAMIC SIMULATION-BASED SOFTWARE

Commonly available dynamic simulation software are Extend[TM][9], ithink[®][10], Simulink[®][11], etc. Dynamic simulation packages typically feature a *flow diagram* interface, enabling modelers to assemble a flow diagram of the system being modeled using graphical icons. The icons contain prepro-grammed "subroutines" and can take one or more inputs, perform a calcula-tion, and produce an output. The icons are assembled in an ordered fashion by the modeler to represent the mathematical model. The flow diagrams are not only mere visual representations of the system being modeled, but are also "active" in that they can simulate the system based on the underlying mathematical model encoded.

The icons that are used to build the flow diagram are comparable, in a way, to the *cells* in the spreadsheet programs. Whereas the links between the cells are "abstract" and not normally visible in the spreadsheets, the links between the icons in the dynamic packages are "physical" and visible. (It is possible to show the links between cells in Excel[®], for example, by turning on the *Trace Precedents* and *Trace Dependents* feature through the *Tools > Auditing* menu.) Spreadsheets, however, represent a *snapshot* of a system, while dynamic simulation packages provide an equivalent of a *moving picture*. The following three dynamic simulation packages are illustrated in this book: Extend[TM], ithink[®], and Simulink[®]. Specific features of these three packages are outlined next.

### 7.4.1 Extend[TM]

The capability of Extend[TM] to handle dynamic models enables problems involving time-based variations of the inputs be modeled with ease. Extend[TM] can handle discrete or continuous variables and linear or nonlinear systems. It has a built-in library of programmed subroutines in the form of icons that can take one or more inputs, perform some calculation, and produce an output. The icons are provided with input and output connectors, and the model is con-structed in the form of a block diagram, by interconnecting icons in an ordered fashion. The developer can also build custom icons with custom functions by

---

[9]Extend[TM] is trademarked by Imagine That, Inc. All rights reserved.
[10]ithink[®] is a registerd trademark of High Performance Systems, Inc. All rights reserved.
[11]Simulink[®] is a registered trademark of The MathWorks, Inc. All rights reserved.

typing in the equations in algebraic form and feeding in inputs from the input connectors to produce a desired output at the output terminal of the icon.

Some of the built-in features of Extend™ that make it ideal for modeling and simulation are animation, plotting, customizable graphical interface, sensitivity analysis, and optimization. Extend™ allows models to be developed in a modular and stepwise manner, whereby users can begin with simple blocks and add features as they learn more about the problem. Different groups of users can develop separate blocks and assemble their blocks effortlessly to complete the model. Automatic dimensional consistency is not maintained by ithink®.

### 7.4.2 ithink®

While ithink® also features a flow diagram interface, only four basic building blocks are used: *stocks, flows, converters,* and *connectors. Stock* blocks, represented by rectangles in the flow diagram and functioning as reservoirs, are accumulators that keep track of the state values at any instant in time. *Flow* blocks, represented by a pipeline with a spigot, let material flow into or out of the stocks at rates specified at the *spigots. Converters,* represented by circles, function as modifiers of flows or containers for model parameters. In a way, stocks can be considered the nouns in the ithink® modeling language, flows are the verbs, and converters are the adverbs. The *connectors,* represented by arrows, link the other three blocks according to the system logic, serving to transmit information (not material) between them.

The converters can receive one or more inputs and generate an output by performing a calculation. The calculation is entered into the dialog box for the converter in the form of an algebraic equation, just as is done in the cell in the spreadsheet-based programs. Once the blocks are connected according to the program logic, and the program is run, ithink® performs a "material balance" across each stock at every time step to update all the state values. The ithink® package also includes built-in animation, plotting, customizable graphical interface, sensitivity analysis, and optimization features. Automatic dimensional consistency is not maintained by the program.

### 7.4.3 Simulink®

Simulink® is another flow diagram-based simulation package for modeling dynamic systems. Simulink® is driven by the mathematical equation solver-based package, MATLAB®. It supports linear as well as nonlinear systems modeled in continuous time or discrete time or a hybrid of the two. It has a unique feature of multirating, i.e., different parts of a system are sampled or updated at different times, if necessary.

The flow diagram in Simulink® is constructed using *Blocks* as in Extend™, by dragging from a library of icons representing sources, sinks, math functions, linear and nonlinear components, connectors, and plotters. Simulink® also allows custom functions to be created through MATLAB® code. Once the system is represented by the flow diagram, it can be run using a choice of integration methods, interactively through the Simulink® menu or, in batch mode, from the command window from MATLAB®. The results can be plotted in Simulink® or put in the MATLAB® workspace for post-processing and visualization.

Simulink® has access to the eight built-in solvers of MATLAB® such as Runge-Kutta method, trapezoidal method, etc., with variable steps and fixed steps, for solving dynamic problems involving differential equations. Higher-order differential equations have to first be reduced to first order by substitution.

## 7.4.4 Extend™ vs. ithink® vs. Simulink®

While these three dynamic simulation packages feature a similar flow diagram-based interface and are functionally almost identical, the following differences may be noted. The ithink® package is more economical in terms of the number of icons required for a model. This is due to the fact that the *Converters* can hold model parameters, such as constants or time-dependent parameters in tabular and/or functional forms, as well as equations to process inputs. In Extend™ and Simulink®, separate icons are required to hold constants, tabular or functional, and time-dependent parameters. Further, when acting as equation holders, *Converters* in ithink® can accept any number of inputs, whereas corresponding icons in Extend™ and Simulink®, accept only a limited number of parameters at their input terminals. If an equation, for example, has to multiply six variables to generate an output, a total of seven *Converters* would be adequate in ithink®—six for holding the inputs and one for multiplying. Extend™ would require six *Constant* blocks, an *Equation* block that has a maximum of five input terminals, and a further *Multiplier* block with two input terminals. MATLAB® would require six *Constant* blocks and five *Multiplication* blocks, each with only two input terminals. The multifunctionality of the *Converters* in ithink® makes building and troubleshooting complex models involving several parameters somewhat easier, when compared to Extend™ and Simulink®.

The plotting capabilities of Simulink® and ithink® are limited, while Extend™ allows some degree of customization. The plotting feature of ithink® allows up to five variables to be plotted in the same frame, while Extend™ allows four at a time, and Simulink®'s *Scope* block allows only one variable at a time. However, taking advantage of the close integration with MATLAB®, additional post-processing features of MATLAB® can be accessed in Simulink® models. The output from the *Plotter* block in Extend™ has a useful feature: it presents the plot as well as a table of the underlying numerical values generated during the simulation in one window. Further, when the cursor is placed at a point inside the plot area, Extend™ displays the numerical value of all of the dependent variables in the plot corresponding to the value of the independent variable at the cursor location. This information is live in that as the cursor is moved, all the values are instantly updated. In ithink®, the numerical values generated during a simulation run can be displayed in a window separate from the plot window.

Extend™ and ithink® have a built-in animation feature for its blocks, while Simulink® does not. The icons in ithink® are rectangles for *Stocks* and circles for *Converters* and *Controllers,* the icons in Extend™ take different graphic forms, and Simulink® features icons commonly used in signal processing and electrical engineering. The icons in Simulink® can be flipped or rotated, enabling the flow diagram to be made more clear. All three of these packages allow submodels to be created by grouping related icons into a common block. Extend™ allows further customization of the icons of the submodels to represent the real system being modeled visually as well. The variables in the equations that the model developer enters into the icons can be custom names in the case of ithink® and Simulink®. In Extend™, the name of the terminal is used rather than the name of the variable that is connected to the terminal.

In the following sections, all of the above eight packages are applied to model a common example problem under various scenarios. The objective of this exercise is to demonstrate their features, advantages, and limitations.

## 7.5 COMMON EXAMPLE PROBLEM: WATER QUALITY MODELING IN LAKES

Consider a lake, of certain volume, $V$, with a flow rate of $Q$ leaving the lake. The lake receives a waste load, $W_{(t)}$, where it is consumed by first-order reactions. The objective is to develop a model to relate the resulting concentration, $C$, of the pollutant in the lake to the system variables under a range of loading and initial conditions. This problem can be modeled based on an MB on the pollutant in the lake as discussed in Section 6.3.1. Assuming the lake to be completely mixed and to be of constant volume, this MB statement can be expressed in the form of the following differential equation:

$$V \frac{dC}{dt} = W_{(t)} - V \cdot K \cdot C - Q \cdot C$$

or

$$\frac{dC}{dt} = \left( \frac{W_{(t)}}{V} \right) - \left( K + \frac{Q}{V} \right) C \qquad (7.1)$$

where $K$ is the sum of the first-order rate constants for all the consumptive reactions. The solution to the above ordinary differential equation (ODE) can yield the concentration of the pollutant in the lake, $C$, as a function of time under predefined initial conditions, $C_0$, and forcing waste load functions, $W_{(t)}$. The forcing functions can be some function of time, continuous or discontinuous. Obviously, the nature of $W(t)$ should be known before attempting to solve this ODE. A general solution to the above can be found from the following:

$$C = \frac{e^{-\alpha t}}{V}\int W_{(t)}e^{\alpha t}dt + C_0 e^{-\alpha t} \tag{7.2}$$

where

$$\alpha = K + \frac{Q}{V} = K + \frac{1}{\tau} \tag{7.3}$$

Obviously, the result will depend on the nature of the forcing function, $W_{(t)}$. The simplest case is the steady state solution, with $C_0 = 0$, under a constant forcing function, $W(t) = W_0$, when the result can be found to be the following:

$$C_{ss} = \frac{W_0}{V \cdot \alpha} = \frac{W_0}{V \cdot K + Q} \tag{7.4}$$

where $C_{ss}$ is the concentration of the pollutant in the lake at steady state.

Because the explicit solution for the problem as described by the general equation [Equation (7.1)] will depend on the initial conditions and the forcing function, it is desirable to develop simulation models that can solve the governing equations in a general manner rather than in a problem-specific manner. Some common types of $W_{(t)}$ functions in this problem are as follows:

- constant loading:
  $W_{(t)} = W_0$
- linearly increasing or decreasing load:
  $W_{(t)} = W_0 \pm at$
- exponentially increasing or decreasing load:
  $W_{(t)} = W_0\, e^{\pm \lambda t}$
- step increase or decrease from a background load of $W_0$:
  $W_{(t)} = W_0$ for $t < 0$ and $W_{(t)} = W_0 \pm W$ for $t < 0$
- impulse load of mass, $m$:
  $W_{(t)} = m\delta(t)$, where $\delta(t) = 0$ for $t \neq 0$ and $\int_{-\infty}^{\infty} \delta(t)dt = 1$
- sinusoidal loading:
  $W_{(t)} = W_{ave} + W_0 \sin(\omega t - \theta)$

By combining the above simple functions, several different real-life loading scenarios can be approximated for modeling purposes. Many of these functions are amenable for analytical solutions, while all of them can be analyzed numerically. However, if parameters of this model are also time-dependent, such as the flow rate, $Q$, for example, then the solution has to be found through numerical simulation. As such, it is desirable to select software packages that can be readily adapted for numerical analysis. Those that have built-in routines for numerical analysis will be preferable for building realistic models.

## 7.5.1 LAKE PROBLEM MODELED IN Excel®

The steady state situation is straightforward to model and simulate. Figure 7.2 shows a "graphical" form of this model set up in the Excel® spreadsheet package. The model inputs $Q$, $W_0$, $V$, and $K$ are entered into cells C8, C9, D12, and D13, respectively. Equation (7.4) is embedded in cell D14 for "$C$" inside the lake with appropriate unit conversion factors. In the spreadsheet shown in Figure 7.2, the mouse has been clicked at cell D9, displaying the "one-line" form of Equation (7.4) in the *formula bar* at the top of the worksheet. Excel® allows users to assign custom names to the cells and to refer to the cells by their names instead of the default column-row notation. This feature is illustrated in this example, which makes the model self-explanatory. It can also be of benefit in troubleshooting.

The equations entered into the cells include the proper unit conversion factors to maintain dimensional consistency in the calculations. The equations

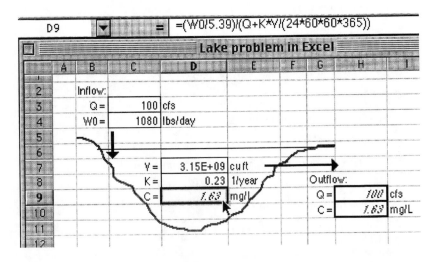

**Figure 7.2** Lake problem modeled in Excel®.

contain references to other cells, which in turn, contain the known input parameters. These cells are referred to by the column-row notation. Alternatively, cells can be assigned appropriate names through the *Insert > Name > Define* . . . menu sequence and can be referred to by their names instead of by the generic column-row notation. This feature may be convenient when building and troubleshooting large models in Excel®.

Several simple yet important scenarios may be simulated by this model. For example, one can vary the $Q$ and $V$ values to understand the impact of the detention time on the in-lake steady state concentration or compare the impact of conservative substances (by setting $K = 0$) against reactive ones ($K \neq 0$). This same lake problem can now be posed in different ways: for example, one way would be to determine the allowable pollutant load, $W$, that would not violate a given in-lake concentration as stipulated, for example, by the water quality standards for the lake. In such a case, this basic spreadsheet model will not be able to back-calculate the answer instantly unless a trial-and-error approach or a built-in function such as goal seeking is adapted.

This form of the spreadsheet model cannot be generalized for different $W_{(t)}$ functions, because a separate model has to be formulated depending on the forcing function. The corresponding algebraic solution must be known in explicit form before constructing the model. To model this lake problem in a general manner, to simulate, for example, different types of waste load inputs, numerical approaches have to be adapted.

For instance, consider the case where the lake is initially under pristine conditions, and a step input of $W = 1080$ lbs/day is input to the lake at time $t = 0$. After five years, this load is reduced instantaneously to $W = 500$ lbs/day. To solve the governing ODE under the above conditions, Euler's numerical method introduced in Chapter 3 has to be used. Figure 7.3 shows such an implementation. The model parameters are defined first. The time step has been chosen as $h = 0.25$ years. (A more accurate solution and a smoother plot can be obtained by using a smaller step such as $h = 0.05$ years.) The waste load input column "W" is filled with the logical expression, if ($t < 5$, 1080, 500), describing the partial step shutdown. By filling the "W" column in this manner, one can easily change the time step $h$ in cell D13 to observe its effect on the final result, without having to adjust the "W" column manually. The last column contains $dc/dt$, which, in essence, is the right-hand side of the ODE to be solved. As can be seen from this implementation, spreadsheets enable simple ODEs to be solved relatively easily through numerical methods.

## 7.5.2 LAKE PROBLEM MODELED IN Mathcad®

The Mathcad® model of the lake problem is illustrated in Figure 7.4. In this model, a case in which $W(t)$ is a time-dependent function is illustrated. Because the particular solutions of the governing differential equations take

Model parameters:

| | | t [yr] | C [mg/L] | W [lbs/day] if(t<5,1080,500) | dC/dt [mg/L-yr] {W/V -$\alpha$ * C} |
|---|---|---|---|---|---|
| Q | 100 cfs | 0.0 | 0.000 | 1080 | 2.0061743 |
| V | 3.15E+09 cu ft | 0.3 | 0.502 | 1080 | 1.3887025 |
| K | 0.23 1/yr | 0.5 | 0.849 | 1080 | 0.9612797 |
| | | 0.8 | 1.089 | 1080 | 0.6654116 |
| | | 1.0 | 1.255 | 1080 | 0.4606074 |

Intermediate calculation:

$\alpha$   1.23 1/yr

Time step for calculation:

h   0.25 yr

| t [yr] | C [mg/L] | W [lbs/day] if(t<5,1080,500) | dC/dt [mg/L-yr] {W/V -$\alpha$ * C} |
|---|---|---|---|
| 3.8 | 1.623 | 1080 | 0.0080534 |
| 4.0 | 1.625 | 1080 | 0.0055746 |
| 4.3 | 1.626 | 1080 | 0.0038589 |
| 4.5 | 1.627 | 1080 | 0.0026712 |
| 4.8 | 1.628 | 1080 | 0.0018490 |
| 5.0 | 1.628 | 500 | -1.0761100 |
| 5.3 | 1.359 | 500 | -0.7448987 |
| 5.5 | 1.173 | 500 | -0.5156295 |
| 5.8 | 1.044 | 500 | -0.3569261 |
| 6.0 | 0.955 | 500 | -0.2470694 |

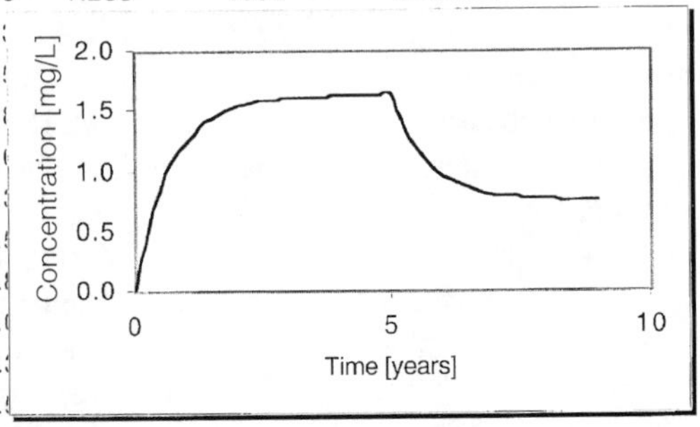

**Figure 7.3** Lake example modeled in Excel® under unsteady state conditions.

different forms for different initial conditions and forcing functions, the complete solution must be known in advance to build models in Excel®. The equation processing engines in mathematical packages such as Mathcad® can solve differential equations in a general manner without the need to modify the basic model. The forcing function in this example is defined to be a partial exponentially declining shutdown from 1080 to 500 lbs/day after six years. In Mathcad®, this partial step shutdown is encoded in a simple statement as follows:

$$W(t) = \text{if}(t > 6 * 365 * 86400, (1080 * e^{-0.005t} + 500), 1080)$$

which is very similar to the format used in the Excel® spreadsheet package for logic statements.

One of Mathcad®'s built-in Runge-Kutta procedures, *rkfixed,* is called to model the problem using the basic form of Equation (7.1). This procedure is

$$V := 3.15 \cdot 10^9 \cdot \text{ft}^3 \quad Q := 100 \cdot \frac{\text{ft}^3}{\text{s}} \quad \mu := -0.05 \cdot \text{yr} \quad Wo := 1080 \quad W1 := 300$$

$$K := 0.00063 \cdot \frac{1}{\text{day}} \qquad W(t) := if\big(t > 5 \cdot \text{yr}, Wo \cdot \exp(\mu t) + W1, Wo\big) \cdot \frac{\text{lb}}{\text{day}}$$

$$C_0 := 0 \qquad\qquad D(t, C) := \left(\frac{W(t)}{V}\right) - \left(K + \frac{Q}{V}\right) \cdot C$$

$$t0 := 0 \qquad N := 100 \qquad tf := 365 \cdot 86400 \cdot 10$$

$$C := rkfixed(C, t0, tf, N, D) \qquad\qquad i := 0, 1 .. N$$

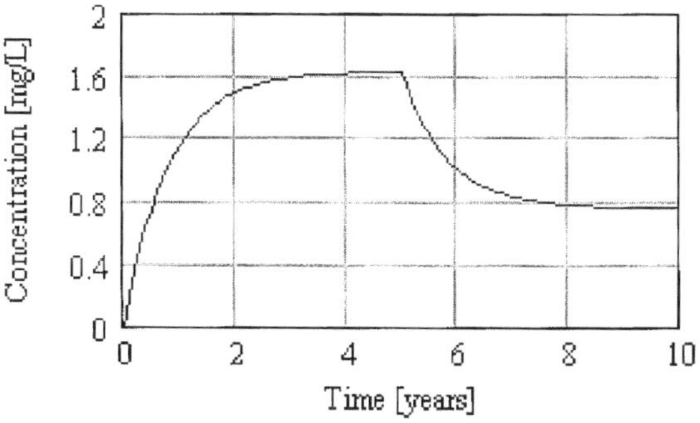

**Figure 7.4** Lake problem modeled in Mathcad®.

based on the fourth-order Runge-Kutta algorithm, which is efficient in solving first-order ODEs. It requires the following arguments to be specified, in order: the dependent variable, the initial value of the independent variable, the final value of the independent variable, the number of points beyond the initial point at which the solution is to be approximated, and the function representing the right-hand side of the ODE to be solved. In the example, these arguments are *C, t0, tf, N,* and *D,* respectively, all of which have to be defined in advance, as shown in Figure 7.4. The procedure returns a matrix, with the independent variable in the first column, ranging from *t0* to *tf,* the corresponding dependent variables in the second column, and the corresponding first derivatives in the third column.

Mathcad®'s built-in ability to perform the calculations to produce results with consistent units is also illustrated in Figure 7.4: the reaction rate constant has been entered in day$^{-1}$ instead of yr$^{-1}$, and the flow rate is entered in cu ft/min instead of cu ft/s. The steady state concentration at $W = 1080$ lbs/day is returned as 1.63 mg/L as before. This feature is fully automatic, saving considerable model building time and avoiding unit conversion errors. The figure also illustrates some of the palettes in Mathcad® that enable mathematical expressions to be displayed in the standard form.

To visualize the results of calculations, another built-in plotting feature is evoked by clicking on the plot button and filling in the placeholders with the appropriate variables for Mathcad® to generate two- or three-dimensional plots. In this example, the results from the Runge-Kutta procedure are used to plot the transient response of the lake as a function of time. The ability to perform the calculations and plot the results in an integrated worksheet provides a clear understanding of the system under dynamic conditions, showing the time to reach steady state condition under a change in $W$.

A unique design feature of Mathcad® is that all the components of the model, such as inputs, constants, equations, etc., are all displayed clearly in the worksheet; thus, the users can "read" the model like a book without needing to know any syntax or opening other sheets, icons, menus, etc., as in other packages. The entire worksheet is live, so that users can experiment with the model by clicking at a value and changing it, letting Mathcad® recalculate and update all the results and plots instantly. Mathcad® solves the equations sequentially by substitution from top to bottom as in traditional programs, and therefore, does not have the ability to backsolve like TK Solver.

## 7.5.3 LAKE PROBLEM MODELED IN Mathematica®

Mathematica®'s ability to solve the governing ODE analytically in symbolic form to yield an analytical solution is illustrated in Figure 7.5. In this example, a constant waste load of $W$ is applied to a pristine lake, starting at time $t = 0$. The built-in procedure *DSolve* is called in line *In[6]* with the equation, initial condition, and the dependent and independent variables in the argument. The algebraic solution is returned in line *Out[6]* as the result. Comparing this result with Equation (7.4) (which was derived using traditional mathematical calculi), it can be seen that they are identical.

The unsteady state lake problem modeled using Mathematica® is shown in Figure 7.6. The model parameters are defined in line *In[27]*. In this case, the waste load is defined as a constant from $t <= 6$ years and as an exponentially declining load for $t > 6$ years. This is specified by a simple logical expression, as in the case of the Excel® spreadsheet example:

$$W = 1080 * 454,000 * 365; \quad \text{waste load} = \text{If}[t > 5, W * e^{-\mu t}, W]$$

In this case, a semicolon is used at the end of the input line to suppress the echoing of the input. A built-in numerical procedure *NDSolve* is called in line *In[30]* to solve the governing ODE. The following arguments are provided for this call: the complete differential equation, the initial value of the

In[6]:= **ClosedForm =**

**DSolve[{c´[t] == $\dfrac{W}{V}$ - $\left(k + \dfrac{q}{V}\right)$ c[t],**

**c[0] == 0}, c, t]**

Out[6]= $\left\{\left\{c \rightarrow \dfrac{W - \dfrac{e^{\left(-k-\frac{q}{V}\right)\#1}\, q\, w}{q+k\,v} - \dfrac{e^{\left(-k-\frac{q}{V}\right)\#1}\, k\, v\, w}{q+k\,v}}{q + k\, v} \;\&\right\}\right\}$

In[7]:= **(c/.ClosedForm[[1,1]])[t]**

Out[7]= $\dfrac{W - \dfrac{e^{t\left(-k-\frac{q}{V}\right)}\, q\, w}{q+k\,v} - \dfrac{e^{t\left(-k-\frac{q}{V}\right)}\, k\, v\, w}{q+k\,v}}{q + k\, v}$

**Figure 7.5** Analytical solution for the lake problem obtained using Mathematica®.

In[27]:= **W=1080.0*454000.0*365.0;K=0.23;**
**wasteload:=If[ t>=5., W*Exp[-μ*t],W];**
**Q=100.0*60 *60 *24 *365; V=3.15*10^9; μ=0.05;**

In[30]:= **NDSolve[{c´[t] == $\dfrac{wasteload}{V*28.32}$ - $\left(\dfrac{Q}{V} + K\right)$ c[t],**
**c[0] == 0.}, c, {t, 0, 10}]**

Out[30]= {{c → InterpolatingFunction[{{0., 10.}}, <>]}}

In[31]:= **Plot[Evaluate[c[t]/.%],{t,0,10}, AxesLabel-**
**>{"Time [yrs]" , "Conc.in lake [mg/L]"},**
**GridLines->Automatic]**

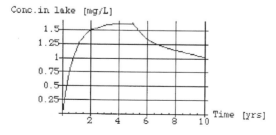

**Figure 7.6** Lake example modeled with Mathematica® under unsteady state conditions.

dependent variable, the dependent variable, and its range over which the ODE is to be solved. The procedure returns the solution in the form of an interpolating function in line *Out[30]*.

Finally, the *Plot* procedure is called in line *In[31]* to plot the results returned by the *NDSolve* procedure. Unlike in Excel® and Mathcad®, where one clicks a button to generate graphs, in Mathematica®, the *Plot* procedure has to be called, specifying all the objects associated with the plot. In the example shown, the arguments indicate the function to be plotted, the range of the independent variable, and the titles for the two axes. Additional plot object specifications can be optionally included with the call to further customize the plot.

## 7.5.4 LAKE PROBLEM MODELED IN MATLAB®

MATLAB®, like Mathematica®, can solve simple ODEs analytically, when the Symbolic Math Toolbox is available. This feature is illustrated first in this example, where a constant waste load of $W$ is applied to a pristine lake, starting at time $t = 0$. The built-in procedure *dsolve* is called with the equation and the initial condition as the arguments. The algebraic solution is returned as the result, as shown in Figure 7.7. This solution was obtained by entering the first line in the *Command* window directly. Comparing this result with Equation (7.4) (which was derived using traditional mathematical calculi), as well as that returned by Mathematica®, it can be seen that all can be reduced to the same form.

The MATLAB® model of the unsteady state lake problem is presented in Figure 7.8. In this case, an exponential declining load of $W = W_0 e^{-\mu t}$ is applied to a pristine lake, starting at time $t = 0$. The objective is to plot the response of the lake for various values of $\mu$. The governing ODE is first set up in an M-File named Lake.m, where a custom function, *Lake*, has been defined as a two-dimensional matrix in line 1. The model parameters, $W$, $Q$, $V$, $K$, and $\mu$ are defined in this M-File to be *global* in line 2, so that they can be interactively changed in the *Command* window without having to edit the M-File that contains the *script*. The governing ODE is entered in line 3.

The model is run from the *Command* window, where the model parameters are first defined to be global. Numeric values for the parameters are then

```
» dsolve('Dc=(w/v)-(k+q/v)*c', 'c(0)=0')

ans =

w/(k*v+q)-exp(-(k*v+q)*t/v)*w/(k*v+q)
```

**Figure 7.7** Analytical solution for the lake problem in MATLAB®.

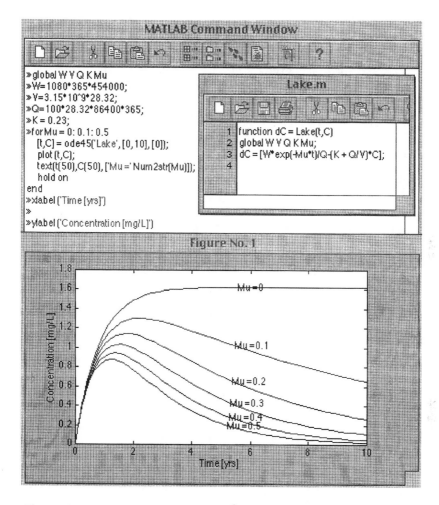

**Figure 7.8** Lake example modeled in MATLAB® for various exponentially decaying loads.

supplied in the following lines. A *For* loop is set up to solve the governing ODE for μ ranging from 0 to 0.5, in steps of 0.1. A built-in function, *ode45*, is used here to solve that ODE. The custom function *Lake* defined in the M-File is used as one of the arguments for *ode45*, which identifies the ODE to be solved. The range of the independent variable over which the solution is sought and the initial value of the dependent variable are also specified by the other two arguments of this call. The result from *ode45* is returned in the form of a matrix [*t, C*], which is then fed to the built-in plot function to generate the graph of *C* vs. *t*. The *text* command is used to add annotations to the plot by specifying the locations and the text to be inserted. The *hold on* command is used to keep the plot frame the same until all the μ values have been evaluated.

As can be seen from this example, several lines of commands have to be typed in the *Command* window every time the program is run. A major part of the commands shown in this example can be moved to the *Lake* M-File or to another M-File to make the program less tedious to maintain and run. MATLAB® also contains a set of tools to construct a graphic user interface to improve the interactivity.

Plotting graphs in MATLAB® also has to be *scripted* with specifications for all the plot objects, such as axis labels, annotations, etc., as in the case of Mathematica®. In addition, unit conversion factors have to be provided by the modeler/user as MATLAB® does not maintain dimensional homogeneity. As can be seen from this example, the MATLAB® environment demands some degree of programming skills, as does the Mathematica® environment.

## 7.5.5 LAKE PROBLEM MODELED IN TK Solver

Figure 7.9 shows the TK Solver model for the steady state solution for the lake problem, illustrating the backsolving feature. The equation, as entered in the *Rule Sheet,* is in the same form as Equation (7.1), but the $C_{ss}$ is entered as a known value in the *Variable Sheet* (= 1.63 mg/L) so as to determine the allowable waste load, $W_0$. When *run,* TK Solver inverts the equation specified in the rule sheet to calculate the unknown $W_0$ and returns the result (= 1081) in the *output* column. Also of interest to note, in TK Solver, all the parameters in the right-hand side of the equation need not be known to solve the model—this is not possible with the other packages.

TK Solver also has the ability to maintain dimensional homogeneity and to perform calculations in a consistent set of units to provide the result in the correct unit; however, this feature is not fully automatic as in the case of Mathcad®. The developer has to incorporate a unit conversion sheet specifying the conversion factors between the units in which the calculations are done and the units in which the variables are displayed in the output sheet. This feature is also illustrated in Figure 7.9, where the reaction rate constant is entered in day$^{-1}$ instead of in yr$^{-1}$; TK Solver automatically calculates and displays the same numerical value as before for the waste load in the units specified in the output sheet.

In contrast to Mathcad®, in TK Solver, the various components of the model, such as inputs, equations, and plots, etc., are displayed in separate windows. Also, the model is not live as in the other packages. After making any changes in the model parameters, the user has to use the menu command *Solve* to recalculate the results or update the plots. TK Solver includes a *MathLook* window, which reformats the line form of the equation entered into the *Rule* sheet, into a standard mathematical form for checking complex equations or for presentation purposes.

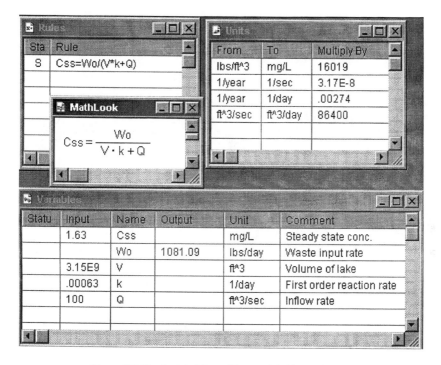

**Figure 7.9** Steady state lake problem modeled in TK Solver.

The ability of TK Solver to solve differential equations in a general manner using built-in functions is illustrated in Figure 7.10. In this example, the response of an initially pristine lake to the input of an exponentially decreasing waste load, $W(t) = We^{-\lambda t}$ is modeled. The governing equation in this case is as follows:

$$\frac{dC}{dt} = \left(\frac{W}{V}\right)e^{-\lambda t} - \left(K + \frac{Q}{V}\right)C$$

and the initial condition is $C = 0$ at $t = 0$. To solve this equation in TK Solver, a built-in numerical procedure is "called." The above equation is fed to the built-in function ODE-BS as the first argument identifying the independent and independent variables, $t$ and $C$, respectively. The limits of $t = 0$ to 3650 days and the lists associated with the dependent and independent variables, $C$ and $t$, are also included in the call. The parametric values, $K$, $V$, $Q$, and $\lambda$, are input by the modeler in the input sheet, and the conversions between the working and display units are specified as before in the units sheet.

The built-in ODE-BS function used in this example follows the Burlsih-Stoer extrapolation algorithm with adaptive step-size control for high-precision numerical solution of first-order, non-stiff, ODEs. Other similar built-in

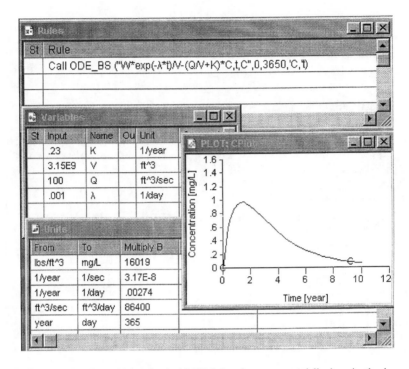

**Figure 7.10** Lake problem solved with TK Solver for an exponentially decaying load.

functions for solving systems of equations and/or stiff ODEs are available in TK Solver.

## 7.5.6 LAKE PROBLEM MODELED IN Extend™

The lake model can be modeled in Extend™ as shown in Figure 7.11. The basic simulation model shown in Figure 7.11 is built using icons from the Extend™ library menu. The *Input data* icon, which can take tabular data, is used in this example to input a partial step shutdown of the waste load as in the Mathcad® example. The *Random input* icon is used to simulate random variations of the flow, $Q$, with a normal distribution of a specified mean and standard deviation. In the example shown, a mean of 100 cfs and a standard deviation of 20 cfs are specified. The three *Constant input* icons are set up for inputting values for $V$, $K$, and the initial concentration in the lake, $Co$.

The *Equation* icon contains Equation (7.3) as entered by the programmer to calculate and return the output as $\alpha$ at the output terminal, using the three inputs, $Q$, $V$, and $K$, fed to the input terminals. Another built-in library routine, *Integrate* icon, for solving the differential equation is used to yield $C$. In

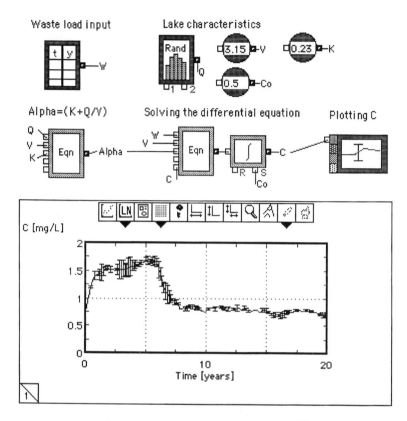

**Figure 7.11** Lake problem modeled in Extend™.

the example shown, Euler's Forward method is chosen with an initial value of $Co = 0.5$ mg/L. Finally, a built-in *Plotter* icon is used to plot the variation of $C$ as a function of time. As shown in Figure 7.11, the error bars indicate the fluctuations that can be expected in $C$ due to random changes in $Q$.

As one can note, the Extend™ flow diagram, on the surface, does not reveal the underlying equations of the model as does the Mathcad® model. Its object-oriented iconic interface is in sharp contrast to the text/numeric appearance of the other applications. Users have to "open" the icons by double-clicking on them to view or edit their contents. However, model parameters and equations are embedded into the icons, just as is done with the other applications using traditional mathematical notations. In fact, the icons in Extend™ are comparable to the cells in a spreadsheet. With advanced skills, the standard icons can be incorporated into more realistic, user-defined graphical objects to present visually meaningful and intuitive models.

The Extend™ package has some advantages over the packages discussed earlier. A major advantage is the program's ability to perform continuous simulation. This enables, for example, models with differential equations to be solved in a general way rather than in a problem-specific manner. Inputs can be random or time variable rather than constant.

## 7.5.7 LAKE PROBLEM MODELED IN ithink®

In ithink®, the model is built in the form of a flow diagram, using the three basic components—stocks, converters, and containers—joined according to the model equations. Based on this flow diagram, the software compiles all the equations and the parameters, as shown in Table 7.1. The flow diagram is shown in Figure 7.12.

This model is also capable of solving the governing differential equation directly, for a time-dependent input of $W$, set by the modeler as a step shutdown function (as was done in the Mathcad® and Extend™ models). This was done simply by entering the following expression into its *container*: Wasteload = IF(Time<6*365) then 1080 else 500. The Euler's method was chosen to solve the differential equation.

The sensitivity analysis feature that is built into ithink® is also illustrated in this model. The sensitivity of the lake concentration to the reaction rate constant ranging from 0.0002 to 0.0008 is set in the sensitivity specification menu for ithink® to generate the concentration profiles at the set $K$ values. Like Extend™, ithink® also generates a table of the calculated values, but only on demand, in a sheet separate from the plot sheet.

## 7.5.8 LAKE PROBLEM MODELED IN Simulink®

The lake problem modeled using the Simulink® package is illustrated in Figure 7.13. In this case, the model is set up so that the waste load = $We^{-\lambda t}$, where $W$ is a step function. The solution of the governing ODE is computed by MATLAB®'s built-in routine, *ode45*, referred to in Section 7.5.4. The initial concentration in the lake can be set in the dialog box for the *Integrator*

Table 7.1 **Lake Model Equations Generated by ithink®**

| | |
|---|---|
| Stock equation: | LakeC(t) = LakeC(t - dt) + (Inflow - RxnLoss - Outflow) * dt<br>INIT LakeC = 0 |
| Inflows: | Inflow = Wasteload*454000/Volume |
| Outflows: | RxnLoss = LakeC*ReactionRate<br>Outflow = Flow_Rate*LakeC/Volume |
| Parameters: | Flow_Rate = 6000*(60*24*28.32)<br>ReactionRate = .00063<br>Volume = 3.15*10^9*(28.32)<br>Wasteload = IF(Time<6*365) then 1080 else 500 |

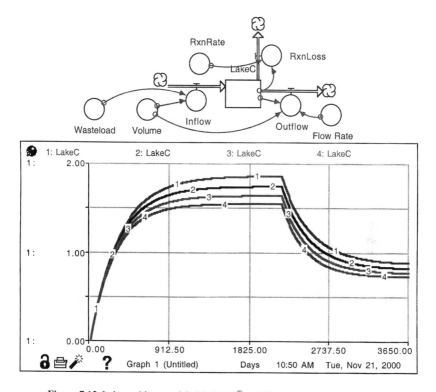

**Figure 7.12** Lake problem modeled in ithink® at different reaction rate constants.

block. The simulation period of 15 years and a time step of 0.1 year are also set through the Simulation Parameters menu.

By setting the value of λ to zero, the same basic model can simulate a step input of *W*. The step function itself can be customized to simulate different magnitudes and/or starting times. By adding other blocks from the Simulink® library, various loading scenarios can be readily simulated. Simulink® also allows parts of the models to be combined into submodels so that complex models can be presented in a compact form.

## 7.6 CLOSURE

As can be seen from the models presented in the last section, the different software packages have their strengths and limitations. In selecting a package for developing simulation models, several factors, such as functional ability, expendability, ease of maintenance, extent of computer skills required, user friendliness, availability, cross-platform transportability, etc., need to be

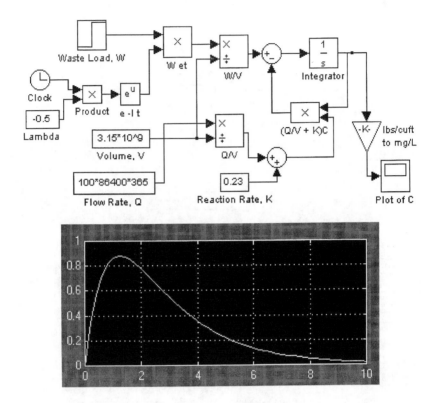

**Figure 7.13** Lake problem modeled in Simulink® for an exponentially declining load.

considered. It is, therefore, impossible and unfair to compare them or rank them or make any specific recommendations.

Some general recommendations based on the above examples and the author's experience are as follows. For those who have not attempted computer-based modeling, Excel® is probably the easiest to learn and to apply to develop models quickly. For those familiar with Excel®, TK Solver should be easy to learn, because the equations are entered into TK Solver using the same one-line style as used in Excel®. Those who have experience in traditional programming should find the environments in Mathematica® and MATLAB® familiar and easy to learn. The learning curve for the dynamic simulation packages is somewhat steep initially, but once some basic understanding is gained, complex models can be developed fairly easily and more quickly than in the other packages.

If the problem is of a static nature, and if the model has to be simulated under various values of the same parameters as inputs without having to rearrange the model equations, then Excel® is probably the most convenient to use. If repeated solutions are required for different combinations of the model parameters by rearranging the model equations, then TK Solver would be more appropriate. Excel® may be used in such cases using the *Goal Seek* and *Solver* options.

Dynamic problems involving a single ODE can be solved with all of the above packages. Mathematica® can find an analytical solution, if possible, and all of them can be used to find numerical solutions. Systems of ODEs can also be solved by all of them; however, the equation solver-based packages and the dynamic simulation packages are more efficient for solving problems involving ODEs. Higher-order equations can be solved directly by Mathematica® and MATLAB®, whereas in Mathcad®, TK Solver, Extend™, ithink®, and Simulink®, they have to be reduced to first order by substitution beforehand. PDEs can be handled efficiently by Mathematica® and MATLAB® but with fairly bulky models in the others. Examples of the use of spreadsheets for solving single, coupled, and partial differential equations have been presented previously (El Shayal, 1990a; El Shayal, 1990b; Kharab, 1988).

Excel®, TK Solver, Extend™, ithink®, and Simulink® packages are data-based, in that numerical values have to be input for the solution. Mathcad®, Mathematica®, and MATLAB® can handle equations symbolically. Mathcad®, Mathematica®, and MATLAB® are best suited for abstract modeling using symbols and for numeric simulations. Mathcad® and Mathematica® feature the "same sheet" interface, where all the inputs, outputs, and interactions are presented in the same screen. The "multiple sheet" environment in MATLAB® demands a steeper learning curve.

Mathcad®, Mathematica®, and MATLAB® feature rich post-processing capabilities for plotting, visualization, animation, and presentation. In comparison, the plotting facilities in Excel®, TK Solver, Extend™, ithink®, and Simulink® serve basic needs, with limited options for customizing. It should also be pointed out that all of the above packages incorporate several other powerful features that are not discussed here, because they are not commonly utilized in modeling environmental systems. For example, spreadsheet packages include statistical tools, database functions, etc., and mathematical packages include curve fitting, complex algebra, etc.

Shacham and Cutlip (1999) have presented a comparison of Excel®, Mathematica®, MATLAB®, Mathcad®, and two other equation solver-type packages, Maple® and POLYMATH, in developing simulation models in chemical engineering. They concluded that all of them were functionally capable and ranked them as follows on the basis of user friendliness and amount of technical effort involved: (1) POLYMATH, (2) Mathcad®, (3) MATLAB®, (4) Mathematica®, (5) Maple®, and (6) Excel®.

---

[12]Maple® is a registered trademark of Waterloo Maple Inc. All right reserved.

It is hoped that the above overview of some of the capabilities, features, and limitations of the different types of software packages will be beneficial to users in selecting appropriate approaches to satisfy their modeling goals. From the wide range of modeling examples included in this book, one can get a general feel for the types of models that can be developed with the different packages. It is probable that with enough practice and familiarization, just about any engineering problem can be modeled with these packages without having to resort to traditional language-based programming, provided the problem can be adequately described mathematically.

As a final note, the quality of many of the plots included in this book may appear poor, because they are "screen shots," made intentionally to illustrate the entire modeling environment. Many of the software packages discussed in this book can, however, print presentation-quality plots directly, if necessary. Some can export high-quality plots to other documents.

## APPENDIX 7.1 SELECTED EXAMPLES OF SOFTWARE PACKAGES SUITABLE FOR MODELING

| Application | Version | Platform | Company | URL |
|---|---|---|---|---|
| *Spreadsheets* | | | | |
| Excel® | Office 98 | PC/Mac | Microsoft Corp. | www./microsoft.com |
| *Equation solvers* | | | | |
| Mathcad® | 2000 | PC | MathSoft Engineering & Education, Inc. | www.mathsoft.com |
| Mathematica® | 3 | PC/Mac | Wolfram Research, Inc. | www.wolfram.com |
| MATLAB® | 5 | PC/Mac | The MathWorks, Inc. | www.mathworks.com |
| TK Solver | 4 | PC | Universal Technical Systems, Inc. | www.uts.com |
| *Dynamic simulators* | | | | |
| Extend™ | 3.2 | PC/Mac | Imagine That, Inc. | www.imaginethatinc.com |
| ithink® | 5.1.1 | PC/Mac | High Performance Systems, Inc. | www.hps-inc.com |
| Simulink® | 3.1 | PC | The MathWorks, Inc. | www.mathworks, com |

# APPENDIX 7.2 EXAMPLES OF TYPES OF EQUATIONS SOLVED BY DIFFERENT SOFTWARE PACKAGES

| | Algebraic Equations | Ordinary Differential Equations | Systems of Ordinary Differential Equations | Higher-Order Differential Equations | Partial Differential Equations |
|---|---|---|---|---|---|
| Excel® | Ex 3-1; Ex 3-2; Ex 5-3; Ex 6-2; §7.5.1; Ex 8-6; Ex 9-11 | Ex 3-4; Ex 3-5; §7.5.1 | | | |
| TK Solver | Ex 8-5; Ex 8-9; Ex 9-6; Ex 9-11 | §7.5.5; Ex 8-4 | | | |
| Mathcad® | §6.3.4; Ex 6-5; Ex 8-1; Ex 9-4; Ex 9-8 | §7.5.2; Ex 8-4 | Ex 8-1; Ex 9-1 | | |
| Mathematica® | Ex 3-1; Ex 3-2; Ex 6-3; Ex 6-4; Ex 6-5; Ex 8-11; Ex 8-12; Ex 9-4; Ex 9-8; Ex 9-12 | Ex 3-3; §7.5.3; Ex 8-4 | Ex 9-1 | | |
| MATLAB® | Ex 3-1; Ex 9-9; Ex 9-10 | §7.5.4; Ex 8-4 | Ex 9-1 | | |
| Extend™ | | §7.5.6; Ex 8-4 | Ex 8-3; Ex 9-1; Ex 9-3; Ex 9-5 | | |
| ithink® | | §7.5.7; Ex 8-4 | Ex 8-1; Ex 8-2; Ex 8-8; Ex 8-10; Ex 9-2; Ex 9-3; Ex 9-7 | Ex 8-7 | Ex 8-12 |
| Simulink® | | §7.5.8; Ex 8-4 | | Ex 8-7 | |

# PART II

# Applications

## CHAPTER 8

# Modeling of Engineered Environmental Systems

## CHAPTER PREVIEW

*In this chapter, 12 examples of engineered systems are illustrated. The selected examples include steady and unsteady state analyses using algebraic and differential equations, solved by analytical, trial and error, and numerical methods. Computer implementation of the mathematical models for the above are presented. The rationale for selecting appropriate software packages for modeling the different problems and their merits and demerits are discussed.*

## 8.1 INTRODUCTION

**T**HIS chapter will serve as the "capstone" chapter for engineered systems in that the principles and philosophies covered in the previous chapters are integrated and applied in the modeling of engineered environmental systems. Engineered systems involving steady and unsteady conditions are modeled applying the general theories presented in Chapter 4 to various reactor configurations discussed in Chapter 5 utilizing the software packages identified in Chapter 7. The first example illustrates the entire modeling process from model development to computer implementation to calibration to validation. The rest of the examples illustrate the model development and computer implementation procedures.

## 8.2 MODELING EXAMPLE: TRANSIENTS IN SEQUENCING BATCH REACTORS

This example is based on a treatability study on a high-strength waste from a soft drink bottling facility, reported by Laughlin et al. (1999). Due to the

**197**

high variation in waste flow rate and its constituents, a sequencing batch reactor (SBR) configuration was chosen to pretreat this waste prior to discharge into the city sewers. This example illustrates the modeling of the SBR process to predict the temporal variation of substrate, dissolved oxygen, and biomass growth during the fill and react phases of the process.

Specifically, the objective of the original study was to investigate whether the bioprocess would be limited by dissolved oxygen levels due to the high substrate concentrations at the beginning of the react phase and to test the hypothesis that an oxygen saturation model can be used to model such limitation. The modeling goal is to achieve a correlation of $r^2 > 0.8$ between predictions and observations of substrate and biomass concentrations at the end of the react phase.

The study consisted of laboratory testing and mathematical modeling. Two bench-scale reactors ("Blue" and "Green") we re-operated in parallel under identical conditions to test reproducibility. These reactors were fed with the high substrate waste from the bottling plant at various concentrations, but the concentrations were maintained constant during each test run. One set of experimental data was used to calibrate the mathematical model, and several other sets were used to validate the model under various flow rates, initial COD concentrations, and operating conditions. Details of the experimental studies can be found in Laughlin et al. (1999).

### 8.2.1 MODEL DEVELOPMENT

Following the classifications in Section 5.22 in Chapter 5, SBRs can be categorized as flow, unsteady during the fill phase, and nonflow, unsteady during the react phase. The significant processes occurring during the fill period are dilution and endogenous decay. The processes occurring during the react period are substrate utilization, microbial growth, endogenous decay, oxygen uptake, and oxygen transfer. Therefore, two sets of material balance (MB) equations have to be developed—one for the fill phase and one for the react phase, with MB equations for biomass, substrate, and dissolved oxygen (DO) for each set. During the fill phase, the only processes occurring are dilution and decay of biomass; hence, the MB equations are as follows:

MB on biomass:

$$\frac{dC_b}{dt} = -\frac{QC_b}{V_0 + QT} - k_d C_b$$

MB on substrate:

$$\frac{dC_s}{dt} = \frac{QC_{s,\text{in}}}{V_0 + Qt} - \frac{QC_s}{V_0 + Qt}$$

MB on dissolved oxygen:

$$\frac{dC_{\text{oxy}}}{dt} = \frac{QC_{\text{oxy, in}}}{V_0 + Qt} - \frac{QC_{\text{oxy}}}{V_0 + Qt} - b'f_d C_b$$

During the react phase, the initial high concentration of the substrate can cause oxygen limiting conditions. To simulate such conditions, a dual Monod's kinetic function is included to modify the biomass growth rate and the substrate utilization rate under low DO levels.

MB on biomass:

$$\frac{dC_b}{dt} = \left(\frac{\mu_{\text{max}} C_s C_b}{K_S + C_s}\right) \times \left(\frac{C_{\text{oxy}}}{K_O + C_{\text{oxy}}}\right) - k_d C_b$$

MB on substrate:

$$\frac{dC_s}{dt} = -\left(\frac{\mu_{\text{max}} C_s C_b}{Y(K_S + C_s)}\right) \times \left(\frac{C_{\text{oxy}}}{K_O + C_{\text{oxy}}}\right)$$

MB on dissolved oxygen:

$$\frac{dC_{\text{oxy}}}{dt} = K_L a(C_{\text{oxy}}^* - C_{\text{oxy}}) - a'\left(\frac{dC_s}{dt}\right) - b'f_d C_b$$

The model parameters were determined from independent experiments, except for the half saturation constant for oxygen $K_o$, which was established during the model calibration process. The variables are defined in Table 8.1.

Table 8.1 Parameters for SBR Model

| Symbol | Definition | Value | Units |
|--------|------------|-------|-------|
| $a'$ | Oxygen-substrate stoichiometric coefficient | 0.2 | mg/mg |
| $b'f_d$ | Respiration rate constant | 0.0075 | 1/hr |
| $C_b$ | Biomass (MLSS) concentration | Variable | mg/L |
| $C_{\text{oxy}}$ | Dissolved oxygen concentration | Variable | mg/L |
| $C_{\text{oxy}}^*$ | Saturated dissolved oxygen concentration | 7.7 | mg/L |
| $k_d$ | Biomass death rate | 0.0004 | 1/hr |
| $K_L a$ | Overall mass transfer coefficient for aeration | 12.8 | 1/hr |
| $K_o$ | Half saturation constant for oxygen uptake | 90 | mg/L |
| $K_s$ | Half saturation constant for substrate | 800 | mg/L |
| $Q$ | Waste flow rate | Variable | L/hr |
| $t$ | Time | Variable | hr |
| $V_0$ | Volume remaining in tank at start of fill phase | 1.8 | L |
| $\mu_{\text{max}}$ | Maximum specific growth rate | 0.2 | 1/hr |

Because the ODEs are coupled, a numerical method has to be used. Further, the inputs are constants; hence, mathematical packages as well as dynamic simulation packages can be used in this example. Models developed with Mathcad® and ithink® are illustrated in the following sections.

### 8.2.1.1 Mathcad® Model

The Mathcad® model is shown in Figure 8.1. The model parameters are first declared in the top section of the sheet. Several parameters are entered using logical statements to switch their values depending on whether the calculations are in the fill phase ($t < tf$) or the react phase ($t > tf$). The syntax of the logic statements in Mathcad® is very similar to that in Excel®.

The differential equations governing the system are entered as a matrix in *D*, which is then fed to the built-in routine, *rkfixed*. The solution is returned as a four-column matrix in *Z*, containing the time, biomass, substrate, and DO concentrations as a function of time. Finally, the fourth column is plotted to show the DO variation. Notice that in this model, the aeration is switched off during the fill period by the following statements: Kla(t) = if(t<tf, 0, 12.8) and μm(t) = if(t<tf, 0, 5.0).

### 8.2.1.2 ithink® Model

The model flow diagram for the SBR developed with ithink® is shown in Figure 8.2. It shows the three separate, but interconnected, model segments, each describing the fate of substrate, biomass, and DO. The *ghosting* feature of ithink® is used here to minimize the complexity of the flow diagram. For example, instead of feeding inputs directly from the substrate *stock* directly to the converters where it is used, ghosts of the substrate *stock* (indicated by dashed lines) are used.

Model parameters, inputs, and intermediate calculations are contained within the *containers* indicated by circles. Components of the MB equations are embedded into the *converters*. All the parameters, inputs, and equations compiled by ithink® are shown in Table 8.2. The initial values for the three *stocks* are set for each *stock* individually. The Runge-Kutta fourth-order method is selected for the three *stocks*.

A user-friendly graphic user interface (GUI) for this model is presented in Figure 8.3. Using the built-in features of ithink®, a GUI is constructed that enables users to adjust several model parameters such as fill time, waste flow rate, and influent COD, interactively. The users can also evaluate the effect of *turning on* aeration during the fill phase.

$tl := 70 \cdot min$    $t := 0 \cdot sec$    $Qin := 4.2 \cdot \dfrac{liter}{hr}$    $Sinit := 700 \dfrac{mg}{liter}$    $Ko := 90 \cdot \dfrac{mg}{liter}$

$tf := 1 \cdot hr$    $fn := .18$    $Cs := 7.7$    $Vstart := 1.8 \cdot liter$

$\mu m(t) := if\left(t < tl, 0, 5.0 \cdot \dfrac{1}{day}\right)$    $Ks := 800 \cdot \dfrac{mg}{liter}$    $Y := .53 \cdot \dfrac{mg}{mg}$    $kd := 0.01 \cdot \dfrac{1}{day}$

$Q(t) := if(t < tf, Qin, 0)$    $b(t) := if\left(t < tf, 0, 0.0075 \cdot \dfrac{1}{hr}\right)$    $So := 240 \cdot \dfrac{mg}{liter}$

$Kla(t) := if\left(t < tf, 0, 12.8 \cdot \dfrac{1}{hr}\right)$    $V(t) := Vstart + if(t < tf, Q(t) \cdot t, Qin \cdot tf)$

$a(t) := if\left(t < tl, 0, .20 \cdot \dfrac{mg}{mg}\right)$    $Sn := \left(fn \cdot 0.7 \cdot So + Sinit \cdot \dfrac{1.8}{6}\right)$    $X := \begin{pmatrix} 5540 \\ 700 \\ .1 \end{pmatrix}$

$$D(t,X) := \begin{bmatrix} if\left[X_1 > Sn, \left[\left(\dfrac{-X_0 \cdot Q(t)}{V(t)}\right) + \dfrac{\mu m(t) \cdot X_0 \cdot X_1}{(Ks + X_1)} \cdot \dfrac{X_2}{Ko + X_2} - kd \cdot X_0\right], -kd \cdot X_0\right] \\ if\left[X_1 > Sn, \left[\dfrac{(So - X_1) \cdot Q(t)}{V(t)} - \dfrac{\mu m(t) \cdot X_0 \cdot X_1}{(Ks + X_1) \cdot Y} \cdot \dfrac{X_2}{Ko + X_2}\right], 0\right] \\ if\left[X_1 > Sn, -a(t) \cdot \left[\dfrac{\mu m(t) \cdot X_0 \cdot X_1}{(Ks + X_1) \cdot Y} \cdot \dfrac{X_2}{Ko + X_2}\right], 0\right] + \left[-b(t) \cdot X_0 + (Kla(t)) \cdot (Cs - X_2)\right] \end{bmatrix}$$

$Z := rkfixed(X, 0, 23400, 156, D)$    $i := 0, 1 .. 156$

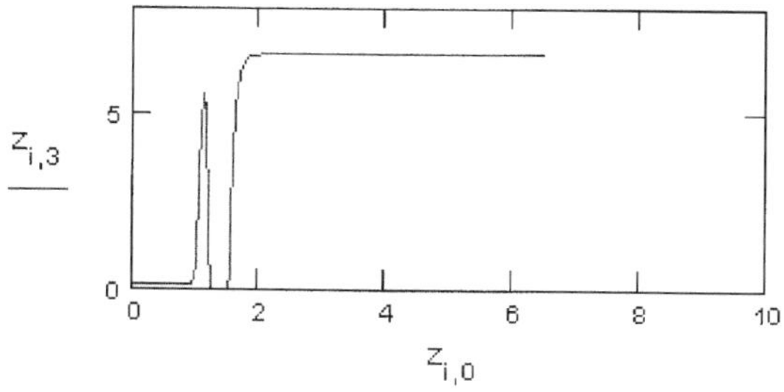

**Figure 8.1** SBR model in Mathcad®.

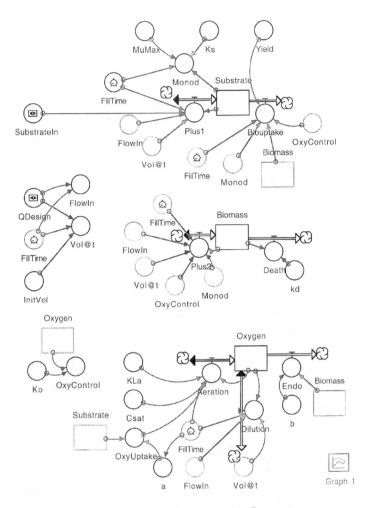

**Figure 8.2** SBR model in ithink®.

## 8.2.2 MODEL CALIBRATION

One set of experimental data was used to establish a value for $K_o$. Starting from literature values, a trial-and-error approach was used to find the optimal value of $K_o = 90$ mg/L. The criterion was to achieve a correlation of $r^2 > 0.8$ between calculated and measured COD and biomass values at the end of the react phase. Some of the biokinetic parameters were also adjusted to be within ±10% of the measured values to improve the degree of fit between calculated and measured COD and MLSS results.

Table 8.2 SBR Model Equations Generated by ithink®

```
STOCK EQUATIONS:
  Biomass(t) = Biomass(t – dt) + (Input2 – Death) * dt
      INIT        Biomass = 5000
      INFLOWS:    Input2 = if(TIME < FillTime) then (-(FlowIn/Vol@t)*Biomass)
                                   else (Monod*OxyControl*Biomass)
      OUTFLOWS: Death = Biomass*kd
  Oxygen(t) = Oxygen(t – dt) + (Aeration – Endo – Dilution) * dt
      INIT        Oxygen = .1
      INFLOWS:    Aeration = if (TIME < FillTime) then 0 else
                                   (Kla*(Csat–Oxygen)–OxyUptake)
      OUTFLOWS: Endo = Biomass*b
                Dilution = if (TIME < FillTime) then (-FlowIn*Oxygen/Vol@t)
                else 0
  Substrate(t) = Substrate(t – dt) + (Iput1 – Biouptake) • dt
      INIT        Substrate = 100
      INFLOWS:    Iput1 = if(TIME < FillTime) then (FlowIn*SubstrateIn/Vol@t)–
                                   (FlowIn*Substrate/Vol@t) else 0
      OUTFLOWS: Biouptake = if (TIME < FillTime) then 0 else
                                   (Monod*Biomass/Yield)*OxyControl
CONSTANTS:  b = 0.0075
            Yield = 0.53
            Csat = 7.7
            QDesign = 3.2
            SubstrateIn = 338
            FillTime = 1
            InitVol = 1.8
            kd = 0.01/24
            KLa = 12
            Ko = 9
            Ks = 800
            MuMax = 5/24
            Monod = if (TIME < FillTime) then 0 else
                                   MuMax*Substrate/(Ks+Substrate)
            OxyControl = Oxygen/(Oxygen+Ko)
            OxyUptake = –a*DERIVN(Substrate,1)
            FlowIn = IF(TIME < FillTime) then QDesign else 0
            Vol@t = InitVol+ (if(TIME < FillTime) then TIME*QDesign else
                FillTime*Qdesign)
            a = if (TIME < FillTime) then 0 else 0.2
```

## 8.2.2.1 Model Validation

The model is validated using the following measures:

(1) Predicted vs. measured concentrations of substrate (COD) and biomass (MLSS) at the end of the react phase
(2) Predicted vs. measured temporal concentration profiles of substrate (COD) and biomass (MLSS) during the fill and react phases

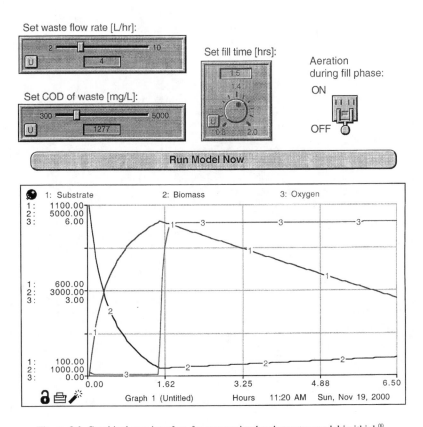

Set waste flow rate [L/hr]:

2 — 10
4

Set COD of waste [mg/L]:

300 — 5000
1277

Set fill time [hrs]:

1.5
1.4
0.8   2.0

Aeration
during fill phase:

ON

OFF

Run Model Now

1: Substrate        2: Biomass        3: Oxygen

Graph 1 (Untitled)    Hours    11:20 AM    Sun, Nov 19, 2000

**Figure 8.3** Graphical user interface for sequencing batch reactor model in ithink[®].

(3) Predicted vs. measured temporal DO profiles during the fill and react phases
(4) Predicted vs. measured long-term substrate (COD) removal efficiencies
(5) MB closure at the end of the react phase to check if the increase in biomass equaled the decrease in substrate times yield

Results of these comparisons are presented in Figures 8.4 to 8.7, which indicate that the model performance is within the expected goals.

## 8.3 MODELING EXAMPLE: CMFRs IN SERIES FOR TOXICITY MANAGEMENT

This example is a modified version adapted from Weber and DiGiano (1996). An industry is considering equipping an existing tank (of volume $V$)

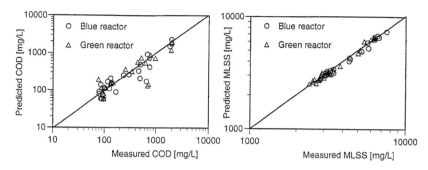

**Figure 8.4** Predicted vs. measured COD and MLSS after the react phase.

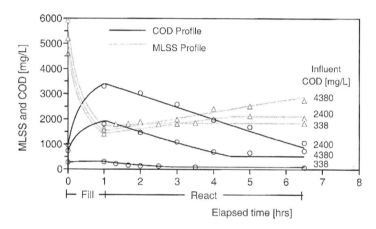

**Figure 8.5** Predicted vs. measured COD and MLSS during the fill and react phases.

as an activated sludge pretreatment system for treating their waste stream to meet the sewer discharge permit. This pretreatment system is expected to receive BOD and a nonbiodegradable, toxic chemical according to the schedule shown below, every two days:

| | 0:00 to 9:00 | 9:00 to 10:00 | 10:00 to 11:00 | 11:00 to 12:00 | 12:00 to 13:00 | 13:00 to 14:00 | 14:00 to 15:00 | 16:00 to 0:00 |
|---|---|---|---|---|---|---|---|---|
| BOD conc., $C$ (mg/L) | 50 | 60 | 75 | 100 | 125 | 142 | 150 | 50 |
| Toxicant conc., $C_T$ (mg/L) | 0 | 30 | 45 | 50 | 50 | 50 | 40 | 0 |

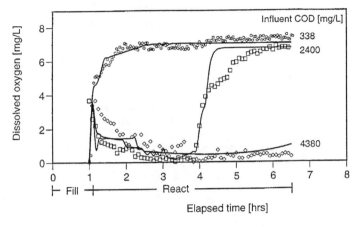

**Figure 8.6** Predicted vs. measured dissolved oxygen at various influent COD levels.

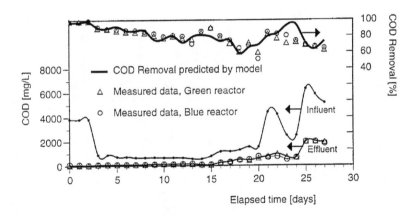

**Figure 8.7** Predicted vs. measured long-term COD removal.

It has been found that if the concentration of the toxicant, $C_T$, exceeds 25 mg/L in the bioreactor, it will be toxic to the primary strains in activated sludge. Additional available data are as follows: hydraulic retention times of existing tank = 0.6 day and biodegradation rate constant, $k$, assumed first order, = 3 day$^{-1}$ provided $C_T$ in the tank is less than 25 mg/L. The sewer discharge permit specifies a nominal limit of 20 mg/L BOD and a maximum limit of 30 mg/L BOD.

It is required to evaluate the feasibility of using the existing tank as an activated sludge unit to treat the waste to the required levels and to make recommendations to the industry. This problem is best solved by developing a mathematical model. First, a process model is developed based on the existing conditions:

MB on toxicant:

$$V \frac{dC_T}{dt} = QC_{T,\text{in}} - QC_T$$

$$\frac{dC_T}{dt} = \frac{1}{\tau}(C_{T,\text{in}} - C_T)$$

MB on BOD:

$$V \frac{dC}{dt} = QC_{\text{in}} - QC - kVC$$

$$\frac{dC}{dt} = \frac{1}{\tau}C_{\text{in}} - \frac{1}{\tau}C - kC = \frac{1}{\tau}C_{\text{in}} - \alpha C$$

where $\alpha = (1/\tau + k)$; and the reaction rate constant $k$ is defined as follows:

$$k = 3 \text{ day}^{-1} \text{ if } C_T < 25 \text{ mg/L and } k = 0 \text{ if } C_T > 25 \text{ mg/L}$$

The above MB equations are simple ODEs that can be solved analytically and implemented in a spreadsheet if the inputs are either constants or simple functions of time. Equation solving and mathematical packages can be used if the inputs can be expressed as mathematical functions. In this example, the influent concentrations $C$ and $C_T$ are arbitrary functions of time as shown in the schedule above; hence, numerical methods have to be used for solving them. The dynamic software packages, Extend™ and ithink®, have all the built-in features for solving this problem. In the next section, models developed with ithink® are illustrated.

Figure 8.8 shows the preliminary model, where the two input variables $C$ and $CT$ are entered as tables into the *containers* labeled "CtoxicantIn" and "CBODin." The symbol ~ inside the circles representing these containers indicates that they contain tabular values instead of the normal constant values. Data from each row in these tables are used to update those variables at each step of the calculation. Two *stocks* are used to represent the toxicant concentration and the BOD concentration. The *container* labeled RxnRate contains the reaction rate expressed as RxnRate = if (Toxicant<25) then 3/24 else 0, where the value for "Toxicant" is drawn from the stock labeled "Toxicant" at each step of the calculation. The model flow diagram is easily constructed using the *connectors*. As the *containers* and *stocks* are connected and filled in, ithink® automatically compiles all the equations to be used in the calculations, as shown in Table 8.3.

When the model is run, ithink® can either generate a table of the output values for every step of the calculation or plot them as a function of time, as

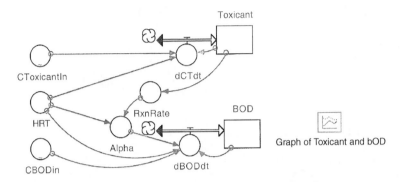

**Figure 8.8** Pretreatment systems: preliminary model using ithink®.

shown in Figure 8.9. From the toxicant and BOD concentration profiles shown, it can be noted that the system as proposed can maintain the toxicant concentration well below 30 mg/L without inhibiting the bioprocess in the reactor; however, the effluent BOD concentration will be in excess of the discharge permit for over 12 hours of the day.

One option for meeting the discharge permit is to install a baffle to partition the tank to form two CMFRs in series. (Refer to Section 5.23 of Chapter 5,

Table 8.3 Pretreatment Model Equations Generated by ithink®

STOCK EQUATIONS:  Toxicant(t) = Toxicant(t − dt) + (dCTdt) * dt
                 INIT        Toxicant = 8
                 INFLOWS:  dBODdt = (CBODin/HRT)−Alpha*BOD
                 BOD(t) = BOD(t − dt) + (dBODdt) * dt
                 INIT        BOD = 20
                 INFLOWS:  dCTdt = (CToxicantIn−Toxicant)/HRT

EQUATIONS:       Alpha = RxnRate+1/HRT

CONSTANTS:       HRT = 0.6*24
                 RxnRate = if (Toxicant<25) then 3/24 else 0

CBODin = GRAPH(TIME)
   (0.00, 50.0), (1.00, 50.0), (2.00, 50.0), (3.00, 50.0), (4.00, 50.0) (5.00, 50.0),
   (6.00, 50.0), (7.00, 50.0), (8.00, 50.0), (9.00, 60.0), (10.0, 75.0), (11.0,
   100), (12.0, 125), (13.0, 142), (14.0, 150) (15.0, 50.0), (16.0, 50.0), (17.0,
   50.0), (18.0, 50.0), (19.0, 50.0), (20.0, 50.0), (21.0, 50.0), (22.0, 50.0),
   (23.0, 50.0), (24.0, 50.0)

CToxicantIn = GRAPH(TIME)
   (0.00, 0.00), (1.00, 0.00), (2.00, 0.00), (3.00, 0.00), (4.00, 0.00), (5.00,
   0.00), (6.00, 0.00), (7.00, 0.00), (8.00, 0.00), (9.00, 30.0), (10.0, 45.0)
   (11.0, 50.0), (12.0, 50.0), (13.0, 50.0), (14.0, 40.0), (15.0, 0.00), (16.0,
   0.00), (17.0, 0.00), (18.0, 0.00), (19.0, 0.00), (20.0, 0.00), (21.0, 0.00),
   (22.0, 0.00), (23.0, 0.00), (24.0, 0.00)

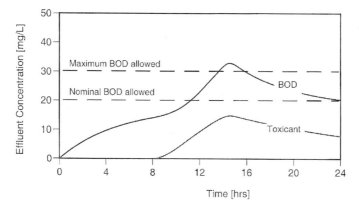

**Figure 8.9** Pretreatment system—effluent profile.

where it was shown that multiple CMFRs in series are more efficient than a single one.) However, because the decrease in toxicant concentration is due only to dilution (or washout), smaller tank volumes may result in inhibitory levels of the toxicant in one or both tanks. An optimal split of the total volume of the tank should, therefore, be found so that both criteria are adequately met. Or, in other words, adequate hydraulic detention time should be allowed in each tank so that the reaction can proceed to reduce the BOD concentration without being inhibited by the toxicant concentration. This involves some degree of trial and error, which can be done efficiently with a modified computer-based model.

The MB equations can be readily derived as follows, with subscript 1 for the first reactor and subscript 2 for the second reactor:

MB equations for toxicant:

$$\frac{dC_{T,1}}{dt} = \frac{1}{\tau_1}(C_{T,\text{in},1} - C_{T,1})$$

and

$$\frac{dC_{T,2}}{dt} = \frac{1}{\tau_2}(C_{T,1} - C_{T,2})$$

MB equations for BOD:

$$\frac{dC_1}{dt} = \frac{1}{\tau_1}C_{\text{in},1} - \alpha_1 C_1$$

and

$$\frac{dC_2}{dt} = \frac{1}{\tau_2}C_1 - \alpha_2 C_2$$

Now, the equations are coupled ODEs with arbitrary inputs, and again, a numerical method has to be adapted for their solution. The computer implementation of this problem using ithink® is shown in Figure 8.10. Even though this diagram can be compacted to be visually more pleasing, it is presented with all the details to illustrate its development.

Sample plots from the trial-and-error process in the search for the optimal split of the volume are shown in Figure 8.11 for a fraction of the total volume in the first tank of 0.40, 0.60, and 0.80. As can be seen from these plots, when the first tank has a volume of 40% of the total volume, the toxicant concentration in that tank exceeds the limit of 30 mg/L, thus resulting in inhibition of the bioprocess and increase in effluent BOD.

The final plots of toxicant and BOD concentrations in the two tanks with 55% of the total volume in the first tank, when all the criteria are adequately met, are shown in Figure 8.12.

In the ithink® package, the modeler can build a GUI so that users can run the model under various conditions without having to interact directly at the modeling layer. One such interface is shown in Figure 8.13 with a *slider* input device, where users can set a volume fraction (within a range preset by the modeler) and click on the *Run Model* button to plot the effluent profiles from each tank. A useful feature of the slider is that it lets users change the value at any time during the run, interactively, and update all calculations thereafter (however, in this example, that feature is of no practical use). Users can also click on the two icons depicting the arbitrary toxicant and BOD input profiles and modify them to simulate other scenarios.

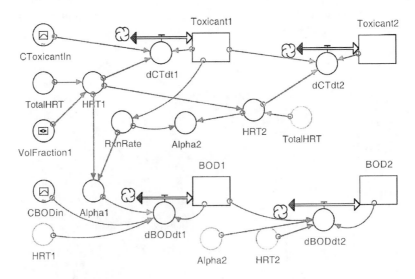

**Figure 8.10** Pretreatment system—ithink® model of modified system.

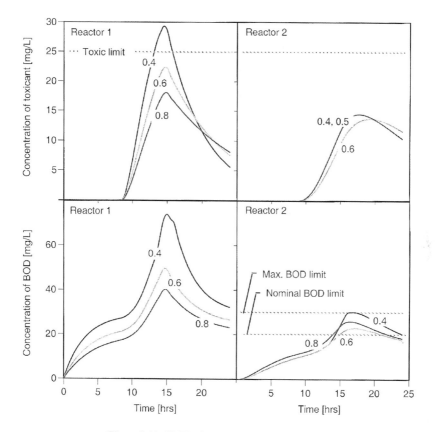

**Figure 8.11** CMFRs in series: optimization study.

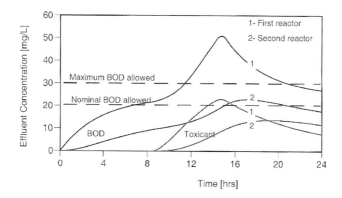

**Figure 8.12** Pretreatment system: concentration profiles under optimized conditions.

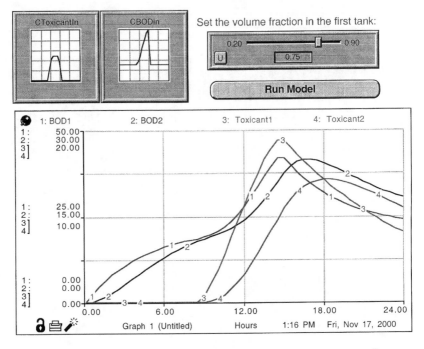

**Figure 8.13** Graphical user interface model for pretreatment system in ithink®.

This example illustrates the benefits of a mathematical model as well as the use of dynamic simulation software. The computer implementation of the model enables one to evaluate a rather complex problem easily and fairly accurately under realistic conditions. The model can also be used to conduct sensitivity analyses and to simulate and evaluate other extreme scenarios that may not be feasible with pilot-scale or laboratory tests. Further refinements can be added to this basic model when the users gain a better understanding of the system and the underlying science.

For example, the inhibitory effect of the toxicant in the above evaluation is modeled as an instantaneous shutdown of the reaction beyond a threshold toxicant concentration level, $C_{T,t} = 25$ mg/L, by the following expression:

$$k = 3 \text{ day}^{-1} \text{ if } C_T < 25 \text{ mg/L}; \text{ else } k = 0$$

In reality, this effect might be a gradual increase beyond $C_{T,t} = 25$ mg/L, increasing with an increase of $C_T$. Such an effect can be readily incorporated into the above model by an expression such as the following, for example:

$$k = 3 \text{ day}^{-1} \text{ if } C_T < 25 \text{ mg/L}; \text{ else } k = 3\left(\frac{K_I}{C_T}\right)$$

Other improvements can include alternate total hydraulic retention times, temperature effect on reaction rate constant, Monod-type reactions for BOD removal, oxygen uptake, stripping of volatile chemicals, etc.

## 8.4 MODELING EXAMPLE: MUNICIPAL WASTEWATER TREATMENT

In this example, development of a model for studying the performance of a wastewater treatment plant under dynamic conditions is illustrated. A municipality is evaluating the impact of rapid growth of the city and the consequent increase in the wastewater flow. The plant consists of a primary settling tank, a trickling filter, and a conventional activated sludge unit for BOD removal. Currently, the plant effluent exceeds the BOD limit under peak flow conditions. As an interim measure, it is proposed that a flow equalization tank be added to the existing plant to meet BOD discharge permits under peak flow conditions.

Wastewater treatment plant design and analysis in most textbooks and design manuals are based on average influent flow and BOD values. Such procedures involve only algebraic equations that can be readily solved analytically. In this case, the primary concern is fluctuations of flow and BOD; hence, a dynamic model has to be developed to aid in the evaluation. However, because the fluctuations in flow and BOD are arbitrary functions of time, analytical solutions cannot be found for the governing equations. Therefore, it is necessary to resort to dynamic simulation of the entire system.

The plant was originally designed based on conventional design equations for the trickling filter and the activated sludge processes. Similar equations for those processes will be adapted in this model. Traditionally, designs of equalization tanks for environmental systems have followed graphical or numerical procedures. A mathematical model suitable for continuous dynamic simulation can be developed, starting from MB on the water and the BOD across the equalization tank. The following assumptions about the equalization tank are made: material flow is through the inlet or the outlet only, the tank is completely mixed, and BOD in the tank does not undergo any reaction. Thus, the MB equations are as follows:

MB on water inside the tank:

$$\frac{dV_{(t)}}{dt} = Q_{\text{in}(t)} - Q_{\text{des}}$$

MB on BOD inside the tank:

$$\frac{d(V_{(t)}C_{(t)})}{dt} = Q_{\text{in}(t)}C_{\text{in}(t)} - Q_{\text{des}}C_{(t)}$$

where $V_{(t)}$ is the volume remaining in the tank at any time; $Q_{in(t)}$ is the volumetric inflow at any time; $Q_{des}$ is the desired, constant flow out of the tank; and $C_{(t)}$ and $C_{in(t)}$ are the instantaneous BOD concentrations in the tank and the influent, respectively.

In this example, the Extend™ dynamic simulation package is chosen to model the system. The model segment in which the above two equations for the equalization tank are implemented is shown in Figure 8.14. This segment receives three inputs (RawQIn, QdesignIn, and RawBODIn) and produces two outputs (VolOut and BODOut). Block 1 inside the segment receives two inputs—Input1 (=RawQin) and Input2 (=QdesignIn)—and executes the equation (entered by the modeler into its dialog box) to produce the result as Result1 = Input1 − Input2. (Note that this result is the right-hand side of the ODE describing the MB equation for water.) This result is fed to Block 2, which performs the numerical integration of its input, to produce its output, Result2, which, in this case, is the volume $V_{(t)}$ remaining in the tank at any time, $t$. The modeler can set the initial value for this integrator in the dialog box and choose the preferred numerical model from Euler Forward, Euler Backward, or Trapezoidal.

Blocks 3, 4, and 5 are assembled in a similar manner to solve the two MB equations. It has to be noted that the inputs and outputs are not necessarily flows of material. They are flows of information or values that the *Blocks* can operate upon, following the equations and settings embedded in them.

The entire plant is modeled with graphic icons representing each process, as shown in Figure 8.15. The icons mimic the "unit operations" concept in that each process receives one or more inputs and produces one or more outputs, functioning as subroutines or submodels. All the model parameters and governing equations are fully embedded within each icon, for example, double clicking on the equalization tank icon will reveal the submodel shown in

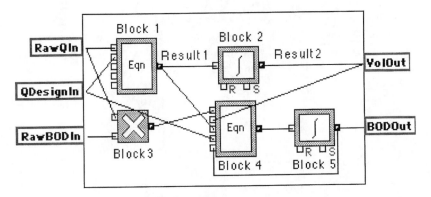

**Figure 8.14** Equalization tank model in Extend™.

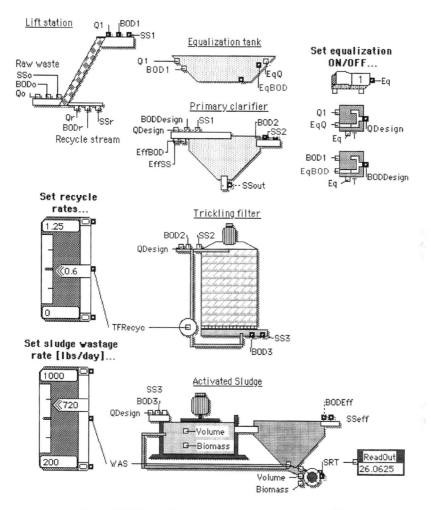

**Figure 8.15** Model of wastewater treatment plant in Extend™.

Figure 8.14 and double clicking on the individual *Blocks* inside the submodel will reveal the governing equations or constants.

The model as presented here includes a simple GUI so that users can interact with the model at a higher level to set model parameters, without having to dive into the model structure. For example, a switch is included that would allow the users to study the system with or without the equalization tank. The output from the switch (= 1 or 0) is fed to two logic boxes that determine the flow rate (EqQ or Q1) and BOD concentrations (EqBOD or BOD1) through the rest of the plant. Similar switches can be added to each process that would allow users to study overall impacts of equipment breakdown, scheduling maintenance shutdown, full load plant capacity, etc.

Sliders are provided so that users can investigate the set different recycle ratios for the trickling filter. Recognizing the fact that adjusting the sludge wastage rate is the primary operational control that is practiced under field conditions, this model includes a slider so that users can try different values. The readout displays the sludge retention time corresponding to the sludge wastage rate. Predictions of effluent BOD with and without the equalization tank from sample runs are shown in Figure 8.16.

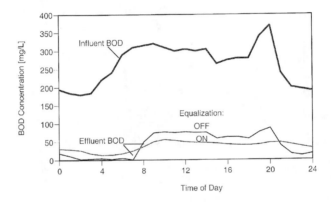

**Figure 8.16a**  Results from wastewater treatment plant model in Extend™ (influent and effluent BOD with and without equalization).

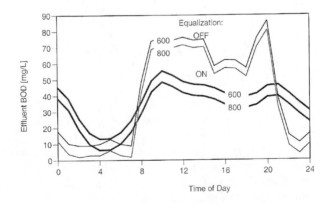

**Figure 8.16b**  Results from wastewater treatment plant model in Extend™ (effluent BOD with and without equalization at sludge wastage rates of 600 and 800 lbs/day).

## 8.5 MODELING EXAMPLE: CHEMICAL OXIDATION

In this example, the development and computer implementation of a mathematical model for a chemical oxidation process are illustrated. This process involves the use of activated carbon as an adsorbent, copper as a catalyst, and oxygen as an oxidizing agent in oxidizing cyanide in industrial wastewaters. The mechanism is surface adsorption and reaction, and the reactor configuration is a CMFR. The operation of the system is as follows: the waste stream flows continuously in and out of the reactor, but the catalyst, activated carbon, is dosed continuously and is retained inside the reactor. Hence, the number of reactive sites inside the reactor will continue to increase, resulting in unsteady conditions. A mathematical model has to be developed to understand and characterize the process.

It can be assumed that the increase in the catalyst due to continuous dosing will not significantly increase the reactor volume, $V$. An MB on cyanide can now be developed as follows:

$$\frac{d(VC)}{dt} = QC_{in} - QC - \eta\{kC\}\{aD(t)\}V$$

$$V\frac{dC}{dt} = QC_{in} - \left[Q - \eta k\left\{a\frac{W(t)}{V}\right\}V\right]C$$

where $C$ is the cyanide concentration in the reactor, $Q$ is the waste flow rate, $\eta$ is the overall effectiveness factor, $k$ is the reaction rate constant, $a$ is the reactive surface area per unit reactor volume, $D(t)$ is the cumulative carbon dose, and $W(t)$ is the cumulative mass of carbon added. Setting $W(t) = DQt$, the MB equation reduces to the following (Weber and DiGiano, 1996):

$$\frac{dC}{dt} = \frac{C_{in}}{\tau} - \left(\frac{C}{\tau}\right)[1 - \eta kat]$$

Even though the final result appears to be simple, it is a nonhomogeneous ODE, and its solution is not straightforward. Procedures to derive the analytical solution to the above can be complicated, and as such, numerical approaches would be appropriate.

Here, Mathematica® is first used to seek an analytical solution. The Mathematica® notebook shown in Figure 8.17 illustrates two different ways in which the analytical solution has been found. In the input line *In(1)*, the built-in function *Dsolve* is called, with arguments that include the equation, the initial condition, the dependent variable, and the independent variable. At this step, the equation is entered in symbolic form, and Mathematica® instantly returns the solution in the output line, *Out(1)*, also in symbolic form, evaluating the integration constant automatically based on the initial

In[1]:= `DSolve[ {c'[t]- (cin - c[t]*(1-const*t))/HRT ==0,`
`c[0]==0},  c[t],t]`

Out[1]= $\left\{\left\{c[t] \rightarrow \dfrac{1}{2\sqrt{const}\ \sqrt{HRT}}\right.\right.$

$\left(e^{-\frac{t}{HRT}+\frac{const\ t^2}{2\ HRT}}\left(cin\ e^{\frac{1}{2\ const\ HRT}}\ \sqrt{2\ \pi}\ Erf\left[\dfrac{1}{\sqrt{2}\ \sqrt{const}\ \sqrt{HRT}}\right]+\right.\right.$

$\left.\left.\left.\left.cin\ e^{\frac{1}{2\ const\ HRT}}\ \sqrt{2\ \pi}\ Erf\left[\dfrac{-1+const\ t}{\sqrt{2}\ \sqrt{const}\ \sqrt{HRT}}\right]\right)\right)\right\}\right\}$

In[2]:= `cin = 40.0;  HRT = 8.0;  const = 2.5;`

`DSolve[ {c'[t]- (cin - c[t]*(1-const*t))/HRT==0,`
`c[0]==0},  c[t],t]`

Out[3]= $\left\{\left\{c[t] \rightarrow e^{-0.125\ t+0.15625\ t^2}\right.\right.$

$\left.\left.(2.03367+11.4938\ Erf[0.0790569\ (-2.+5.\ t)])\right\}\right\}$

In[4]:= `Plot[Evaluate[c[t]/.%],{t,0,5}]`

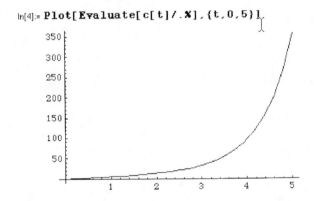

**Figure 8.17** Chemical oxidation process modeled analytically in Mathematica®.

condition provided in the call. If the initial condition is not provided, the solution will include the integration constant c[1].

In the next input line *In(2)*, the model parameters are first assigned numeric values (Cin = 40; HRT = 8; and η.k.a = 2.50 ), and *Dsolve* is called again in symbolic form. Mathematica® returns the solution in the output line *Out(3)*, in numeric form, substituting the given parameter values. Finding the solution in numeric form is a prerequisite for plotting the results. Finally, a call to *Plot* the results is made in the input line *In(3)*, whereupon, Mathematica® plots the results as specified in the call.

When Mathematica® cannot find analytical solutions, it will return a message indicating its inability. In such cases, users can call the Mathematica® built-in numerical procedure, *NDSolve*. This approach is illustrated here to solve the same equation with numeric model parameters. The call to this procedure includes the equation, an initial condition, the dependent variable, and the range of the independent variable over which the numerical solution is found. Of course, in this case, numerical values have to be provided for the model parameters. For the same values of Cin = 40, HRT = 8, and η.k.a = 2.50, the result for *C* is returned as an interpolating function. A call to the *Plot* procedure is then called to plot a graph of the interpolating function (returned by the *NDSolve*) as a function of time, as shown in Figure 8.18.

The same problem can be modeled using MATLAB®. Even though MATLAB® cannot solve this equation analytically, its numerical procedure, *ode45* can, as shown in Figure 8.19. In MATLAB®, an M-File containing the model parameters and the right-hand side of the differential equation is first created and saved under the name of "Oxidation." Then, a call to the numerical procedure *ode45* is evoked from the *command* window, passing the arguments that indicate the name of the M-File containing the differential equation (Oxidation, in this case) and the initial condition. The solution is then plotted using the *plot* call.

The TK Solver package is used to solve this problem as shown in Figure 8.20, where the use of custom functions and advanced built-in procedures to

```
In[13]:= cin = 40.0; hrt = 8.0; const = 2.5;
    NDSolve[{c'[t]- (cin - c[t]*(1-const*t))/hrt
    ==0, c[0]==0}, c, {t,0,5}]

Out[14]= {{c -> InterpolatingFunction[{{0., 5.}}, <>]}}

In[15]:= Plot[Evaluate[c[t]/.%],{t,0,5}, AxesLabel->
    {"Time [hrs]" , "Conc. [mg/L]"}]
```

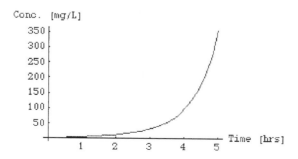

**Figure 8.18** Chemical oxidation process modeled numerically in Mathematica®.

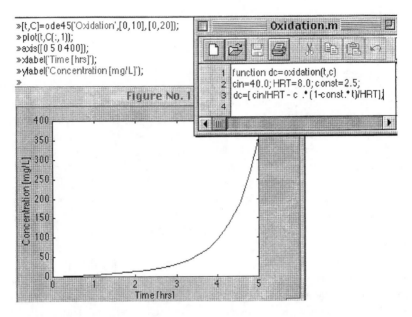

```
»[t,C]=ode45('Oxidation',[0,10],[0,20]);
»plot(t,C(:,1));
»axis([0 5 0 400]);
»xlabel('Time [hrs]');
»ylabel('Concentration [mg/L]');
»
```

Oxidation.m

```
1  function dc=oxidation(t,c)
2  cin=40.0; HRT=8.0; const=2.5;
3  dc=[ cin/HRT - c .* (1-const.*t)/HRT];
4
```

**Figure 8.19** Chemical oxidation process modeled numerically in MATLAB®.

further enhance the functionality of TK Solver is illustrated. First, the built-in Runge-Kutta procedure is evoked in the *Rule* sheet. The first argument to this call, named "Equation" in this example, specifies the equation to be solved, which is entered as a custom procedure. The second and third arguments specify the lists for the dependent variable and the independent variable, $C$ and $t$, respectively, in this example.

Following the *Rule* sheet, the *Functions* sheet automatically lists the built-in procedure, *RK4-se*, and the custom procedure, "Equation." By opening the procedure sheet for "Equation," the governing ODE for this problem is entered in the bottom section, and the parameters and the input and output variables are specified in the top section. The parameters are automatically listed by TK Solver in the *Variables* sheet, where their numerical values can be entered in the *Input* column as usual. A list is then created for the independent variable, $t$, specifying the initial value $(= 0)$, the step $(= 0.1)$, and the final value $(= 5)$. Finally, a list is created for the dependent variable, $C$, and its initial value is also specified $(= 0)$.

When run, the governing equation is fed to the Runge-Kutta procedure to solve the equation over the range of values for $t$ specified in its list and the return solution for $C$ to its list. The *Plot* sheet can be set up to use the $t$-list and the $C$-list to plot a graph of $C$ vs. $t$, as shown in Figure 8.20. Note that the accuracy of the result will depend on the step size specified in the list for $t$.

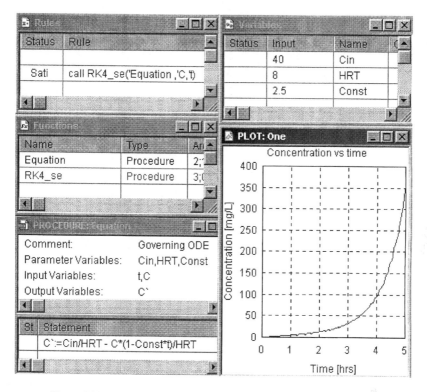

**Figure 8.20** Chemical oxidation process modeled numerically in TK Solver.

The problem is solved in Mathcad® using the built-in numerical routine *rkfixed* as shown in Figure 8.21. The model parameters are first declared at the top of the sheet along with the initial value for the dependent variable, which is defined as a vector, $C_0$. The governing differential equation is then entered using standard mathematical notations that are automatically converted to the two-dimensional form by Mathcad®. The left-hand side of this equation indicates the dependent and independent variables in the equation, $t$ and $C$, respectively. The built-in Runge-Kutta numerical method is evoked by the call *rkfixed*, with the arguments specifying the dependent variable ($C$), the initial and final values for the independent variable (0 and 5), the step for the numerical procedure (100), and the name of the equation to be solved (OxidationODE).

The routine *rkfixed* returns the solution in the 2-column by $i$-row matrix $Z$, the first column $Z^{<0>}$ contains the values of the independent variable, $t$, and the second column $Z^{<1>}$ contains the solution of the equation solved, $C$. The number of rows is equal to the number of steps specified in the call to *rkfixed*.

$C_0 := 0.0$     $K := 2.5$     $HRT := 8$     $Cin := 40.0$

$$Oxi(t, C) := \frac{Cin}{HRT} - C \cdot \frac{(1 - K \cdot t)}{HRT} \qquad Z := rkfixed(C, 0, 5, 100, Oxi)$$

$i := 0 .. rows(Z)$

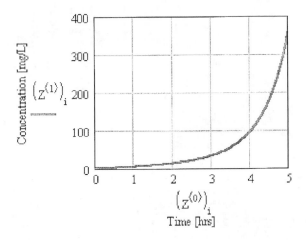

**Figure 8.21** Chemical oxidation process modeled in Mathcad®.

The values contained the matrix $Z$ are then used to plot a curve of $C$ vs. $t$ as shown in Figure 8.21.

With the Pro version of Mathcad®, a simpler model as shown in Figure 8.22 can be formulated to solve problems involving a single ordinary differential equation, such as in this example. Here, the differential equation is entered in the standard form as a *Given* statement, along with the initial condition in a *Solver Block.* Then, a call to the built-in function *Odesolve* is made with arguments specifying the independent variable ($t$), its final value (= 5), and, optionally, the number of steps to be used in the numerical procedure (= 100). The solution is returned as a function of the independent variable, and the results can be plotted directly as shown in Figure 8.22.

This example demonstrates the benefits of the advanced capabilities of equation solving packages that enable users to focus their efforts on model building and simulation aspects rather than on the mechanics of the underlying mathematical calculi or their computer implementation. While this is a definite benefit, it should be noted that these packages are run through commands that require familiarity with their respective syntax and the procedures appropriate to the problem being modeled. For example, each program has its own syntax for calling the same numerical procedure Runge-Kutta, even

Cin := 40        HRT := 8        K := 2.5

Given

$$\frac{d}{dt}C(t) = \frac{Cin}{HRT} - C(t)\cdot\frac{(1 - K\cdot t)}{HRT} \qquad C(0) = 0$$

C := Odesolve(t, 5, 100)

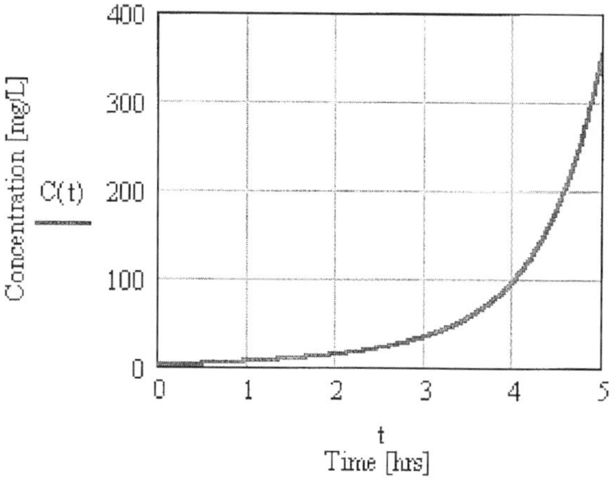

**Figure 8.22** Chemical oxidation process modeled in Mathcad®.

though the procedures for generating the plots are different in each case, with some requiring hard coding and some using a graphical interface. Nevertheless, compared with traditional programming, these packages require minimal syntax knowledge and are more intuitive.

The dynamic simulation program ithink®, for example, enables this problem to be set up readily, as shown in Figure 8.23. The "code" shown as an inset is automatically generated by ithink®, based on the flow diagram assembled using built-in elements—the stock *C*, the flow *dcdt*, and the converters/containers, Cin, HRT, and Constant. While the modeler is expected to know the different syntax for the different routines in MATLAB® and Mathematica®, in ithink®, the procedure for assembling the flow diagram is more or less the same for all types of problems.

The model developed with the dynamic simulation package, Extend™, is shown in Figure 8.24. The model parameters and the initial concentration, *Co,* are stored in *Constant* blocks. The right-hand side of the ODE is formulated in the *Equation* block and fed to the *Integrator* block, which also receives *Co* at terminal S. The solution generated by the *Integrator* block is then plotted by the *Plotter* block.

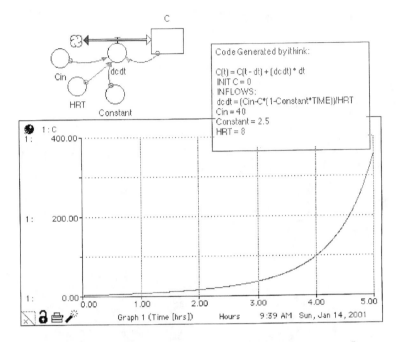

**Figure 8.23** Chemical oxidation process modeled numerically in ithink®.

The model developed using Simulink® and the concentration vs. time plot generated by the *Scope* block in that model are shown in Figure 8.25. In Simulink®, more blocks have to be used to construct the flow diagram to complete the model, because the right-hand side of the ODE has to be formulated with elementary blocks. The *Integrator* block then solves the ODE, and the *Scope* block is used to plot the results.

## 8.6 MODELING EXAMPLE: ANALYSIS OF CATALYTIC BED REACTOR

The process model for a new catalytic fixed-bed reactor being developed for oxidation of pesticides in groundwater is illustrated in this example. Preliminary laboratory studies have indicated that the surface reaction rate (moles/cm²-s) of the bed can be approximated by a first-order process with a reaction rate constant $k$ of 0.01 cm/s. The catalyst is to be coated onto spherical support media of diameters ranging from 0.25–1.5 mm. It is required to evaluate various combinations of process parameters to define optimal ranges for further pilot-scale testing. A mathematical model has to be developed to meet these goals.

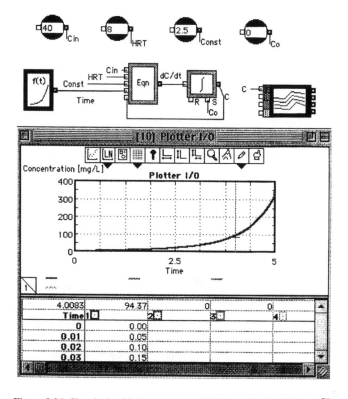

**Figure 8.24** Chemical oxidation process modeled numerically in Extend™.

Equation (5.34) developed in Chapter 5 can be applied here for the bed, assuming dispersive flow with the external resistance controlling the process:

$$\frac{C_{out}}{C_{in}} = \frac{4\beta \exp\left(\dfrac{vL}{2E}\right)}{(1+\beta)^2 \exp\left(\dfrac{vL}{2E}\right) - (1+\beta)^2 \exp\left(-\dfrac{vL}{2E}\right)}$$

where

$$\beta = \sqrt{1 + 4k_f a\left(\frac{E}{vL}\right)}$$

The mass transfer coefficient, $k_f$, can be found from correlations:

$$k_f = \frac{N_{Sh}E}{d_p}$$

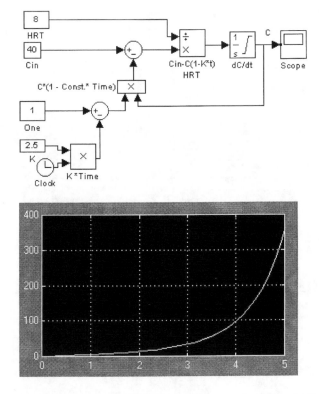

**Figure 8.25** Chemical oxidation process modeled analytically in Simulink[®].

where

$$N_{Sh} = \text{Sherwood No.} = (2 + 0.644 N_{Re}^{0.5} N_{Sh}^{0.333})[1 + 1.5(1 - \varepsilon)]$$

$$N_{Re} = \text{Reynolds No.} = \frac{vd_p\rho}{\mu} = \frac{vd_p}{\upsilon}$$

$$N_{Sc} = \text{Schmidt No.} = \frac{\mu}{\rho E} = \frac{\upsilon}{E}$$

Even though this is a static problem with algebraic equations, Equation (5.34) cannot be solved explicitly to find the bed length for a desired process performance. Webber and DiGiano (1996) recommended a graphical procedure for solving this problem. Further, it would be preferable to be able to solve the equations repeatedly under ranges of input values and under different inputs. Hence, a computer-based model can be of significant benefit. Spreadsheet programs can be used in this case; however, they can become fairly large and may not be suitable for backsolving, if necessary. Equation solving packages are well suited for modeling these types of problems. In this case, TK Solver is selected to develop such a model.

The screen-shot of a portion of the TK Solver model displaying the *Rule* sheet, the *Variables* sheet, and the *Units* sheet is shown in Figure 8.26. It can be noted that the equations can be entered in any order in the *Rule* sheet. The *Units* sheet allows variables to be entered or displayed in customary or convenient units in the *Variables* sheet and, at the same time, perform the calculations in a consistent set of units.

The distinct feature of TK Solver, namely, the ability to *backsolve* Equation (5.34), is illustrated here. The influent concentration and the expected effluent concentration are entered as known values in the *Input* column, along with the other known process parameters. The parameter $L$ is specified as an *Output* in the *Status* column by double clicking at the cell and choosing from a pop-up menu, which is normally not visible. The conversion factors for the display units and the calculation units are specified in the units sheet. When run, TK Solver solves the equations iteratively to return the required length, $L$, that satisfies the specified removal. To initiate an iterative solution process, one of the unknowns has to be assigned an arbitrary *Guess* value as an input

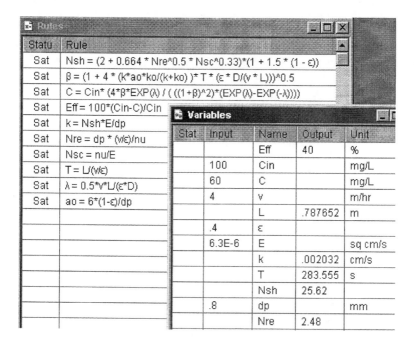

**Figure 8.26** Catalyst process modeled in TK Solver.

in the *Input* column. In this example, a *Guess* value was input for the unknown *L*. On completion of the iteration process, TK Solver returns the correct answer for *L* in the *Output* column.

## 8.7 MODELING EXAMPLE: WASTE MANAGEMENT

In this example, a simple model is developed for multipurpose use in a metal plating facility. The objective of the model is to simulate the facility under various scenarios for use in process analysis, product selection, material balances, record keeping, optimization, waste minimization studies, life cycle analysis, etc. The processes to be included are paint formulation, spray painting off-gas treatment, and recovery of solvents. The model is based on simple material balances using average flows and capacities. The following factors are to be determined for each scenario examined: the best alkyd:epoxy:thinner proportion, the volumes of the three components consumed per year (gal/yr), the concentration (ppmv) of toluene in the off-gases from the painting operation, the amount of toluene (lbs/yr) that can be recovered by using the carbon adsorption system, the concentration (ppmv) of toluene in the off-gases after treatment by activated carbon, the atmospheric emissions (lbs/day) of toluene after installing the activated carbon system, the concentration (mg/L) of toluene in the condensate, and the energy (KW) requirement for the activated carbon system.

Because the model is expected to simulate the system under average conditions, a steady state model would be adequate, and as such, the Excel® spreadsheet package is chosen in this case. The spreadsheet model is built with a graphic interface as shown in Figure 8.27. The model parameters are readily accessible for any changes to be made for evaluation. The appropriate MB equations are embedded in the cells to calculate the model outputs. It should be noted that the spreadsheet is built from "top-down," and specific parameters must be known in advance so that the model can run through the calculations using the known data and calculating the unknowns.

## 8.8 MODELING EXAMPLE:  ACTIVATED
## CARBON TREATMENT

Powdered activated carbon (PAC) is a commonly used adsorbent for treating a wide range of wastewater. In this example, use of PAC in treating a dye bath effluent is modeled to evaluate alternative process configurations. In an existing process, the effluent is contacted with PAC in a CMFR to remove color, and the PAC is recycled after off-line biological regeneration. The process was designed based on the experimental observation that interparticle

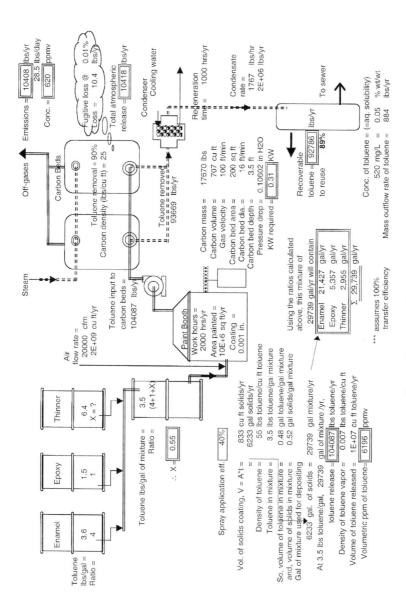

**Figure 8.27** Model of metal plating facility.

diffusion would be controlling the adsorption process. Under continuous operating conditions, however, it was found that the residual biofilm on the PAC surface was causing additional resistance to adsorption in the liquid phase, resulting in lower removals.

The objective here is to develop a model to test the hypothesis that by installing baffles to convert the CMFR to a packed bed PFR configuration with the same detention time would be a cost-effective alternative. The model development for the PFR configuration follows the steps detailed by Weber and DiGiano (1996):

MB on color in liquid phase:

$$\frac{dC}{dt} = -v\frac{dC}{dx} + D\frac{d^2C}{dx^2} - \frac{\text{mass of color adsorbed}}{\text{volume of reactor} \times \text{contact time}}$$

The last term in the MB equation can be expressed as follows:

$$= k_f a(C - C_e)$$

where $k_f$ is the mass transfer coefficient; $a$ is the specific external surface area per unit volume of the reactor; $(C - C_e)$ is the driving force, $C_e$ being the concentration that would be in equilibrium with the adsorbed concentration in accordance with an appropriate isotherm. An expression for $a$ can be developed as follows:

$$\frac{\text{Surface area of PAC}}{\text{Volume of liquid}} = \left(\frac{\text{mass of PAC}}{\text{volume of liquid}}\right)\left(\frac{\text{volume of PAC}}{\text{mass of PAC}}\right)$$

$$\times \left(\frac{\text{area of APC}}{\text{volume of PAC}}\right)$$

$$= (D_{\text{pac}})\left(\frac{1}{\rho_c}\right)\left[\frac{(4\pi R_c^2)}{\left(\frac{4}{3}\pi R_c^3\right)}\right] = \frac{3D}{R_c\rho_c}$$

where $D_{\text{pac}}$ is the PAC dose (= mass of carbon/volume of liquid), $R_c$ is the radius of the carbon particles, and $\rho_c$ is the density of PAC.

Thus, MB on color in the liquid phase is:

$$\frac{dC}{dt} = -v\frac{dC}{dx} + D\frac{d^2C}{dx^2} - k_f\left(\frac{3D_{\text{pac}}}{R_c\rho_c}\right)(C - C_e)$$

and the MB on the solid phase is:

$$\frac{dq}{dt} = \frac{3k_f}{R_c\rho_c}(C - C_e)$$

where $q$ is the mass of color adsorbed per unit mass of PAC. Considering steady state conditions, the final equations are as follows:

$$\frac{d^2C}{dx^2} = \frac{v}{D}\frac{dC}{dx} + \left(\frac{3D_{pac}k_f}{DR_c\rho_c}\right)(C - C_e)$$

$$\frac{dq}{dx} = \frac{3k_f}{vR_c\rho_c}(C - C_e)$$

where $C_e$ and $q$ are related to one another through the isotherm such as:

$$q = KC_e^{1/n}$$

The above equations are coupled nonlinear differential equations, the first one being of second order. Numerical methods have to be used to solve them with equation solver-based packages such as Mathematica®, for example. In this case, the use of the ithink® simulation package in solving a second-order equation is illustrated.

When solving differential equations with ithink® (or with Extend™), higher-order equations must be first reduced to first order. This can be achieved in this case by introducing a new variable $U = dC/dx$. With this substitution, the above second-order equation is reduced to a first-order equation, bringing the total number of ODEs to solve to the following three:

$$\frac{dU}{dx} = \frac{v}{D}U + \left(\frac{3D_{pac}k_f}{R_c\rho_c}\right)(C - C_e)$$

$$U = \frac{dC}{dx}$$

$$\frac{dq}{dx} = \frac{3k_f}{vR_c\rho_c}(C - C_e)$$

The three equations and the isotherm equations are implemented in the ithink® flow diagram as shown in Figure 8.28. The model equations compiled by ithink® are included here in Figure 8.28 to illustrate how the new variable $U$ is incorporated into the calculations to handle higher-order differential equations. Plots of $C$ vs. $x$ and $q$ vs. $x$ from a typical run are shown in Figure 8.29.

The problem can be modeled with Simulink® as shown in Figure 8.30. As can be seen from the two implementations, the ithink® model is more compact.

## 8.9 MODELING EXAMPLE: BIOREGENERATION OF ACTIVATED CARBON

The merits and demerits of integrating biological treatment with activated carbon adsorption in a single reactor configuration over separate systems

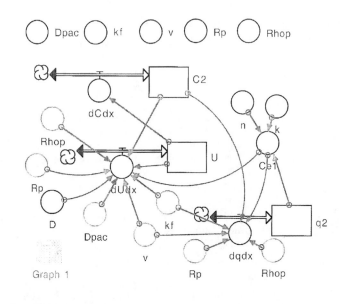

C2(t) = C2(t - dt) + (dCdx) * dt
   INIT:         C2 = 100
   INFLOWS:  dCdx = -U

q2(t) = q2(t - dt) + (dqdx) * dt
   INIT          q2 = 0
   INFLOWS:  dqdx = 3*kf*(C2-Ce1)/(v*Rp*Rhop)

U(t) = U(t - dt) + (dUdx) * dt
   INIT          U = 10
   INFLOWS:  dUdx = U*v/D + 3*Dpac*kf*(C2-Ce1)/(D*Rhop*Rp)

Ce1 = (q2/k)^(1/n)
D = 2.3E-6   Dpac = 1     k = 2.68     kf = 0.072
n = .18       Rp = .001    v = 5       Rhop = 810000

**Figure 8.28** Activated carbon process modeled in ithink®.

have been reported upon recently. As an alternative to such biologically activated carbon (BAC) process, off-line biological regeneration (OBR) of the spent activated carbon has been proposed to alleviate the problems of excessive head loss, short circuiting, and nutrient limitations of the BAC process. In the OBR process, the exhausted packed activated carbon column is taken out of service and regenerated by recirculating a mixture of acclimated biomass, nutrients, and dissolved oxygen. During this regeneration period, the adsorbed chemicals are biodegraded, thereby regenerating the adsorption capacity of the carbon bed.

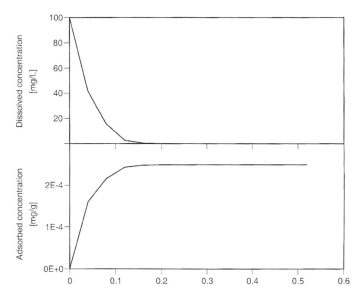

**Figure 8.29** Results of activated carbon model in ithink®.

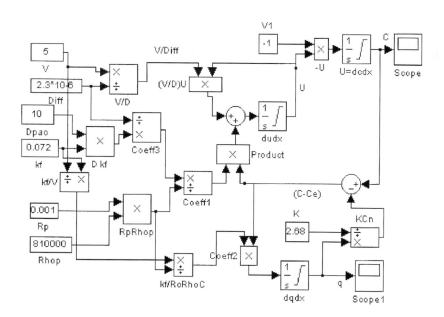

**Figure 8.30** Activated carbon process modeled in Simulink®.

In this example, a mathematical model is developed to describe this process to gain a better understanding of the sensitivity of the process to the various parameters. The following assumptions are made in model development: biological growth is single-substrate limited; biokinetic rates are substrate inhibited according to the Haldane rate expression; the OBR process is completely mixed; and equilibrium conditions exist at the carbon particle surface, according to the Freundlich isotherm. Model development follows the work of Goeddertz et al. (1988). The MB equation for substrate concentration, $C$ (ML$^{-3}$), can be stated as follows:

$$\frac{d(VC)}{dt} = (\text{biodegradation rate}) + (\text{desorption rate})$$

$$\frac{dC}{dt} = -\left(\frac{\mu}{Y}\right)X + \left[\frac{k_d a_s m}{V}\right](C_s - C)$$

where

$$\mu = \mu_{max}\left(\frac{C}{K_s + C + \frac{C^2}{K_1}}\right)$$

$V$ is the volume of the bulk liquid phase (L$^3$), $X$ is the biomass concentration (ML$^{-3}$), $Y$ is the biomass yield (–), $k_d$ is the desorption rate constant (LT$^{-1}$), $a_s$ is the specific surface area of the carbon (L$^2$M$^{-1}$), $m$ is the mass of carbon (M), and $C_s$ is the substrate concentration at the carbon-liquid interface (ML$^{-3}$). It is convenient to lump the group of variables ($k_d a_s m/V$) for measurement and computation purposes. Using the Freundlich isotherm relationship, $C_s$ can be expressed as follows:

$$C_s = \left[\frac{1}{k_f}\left(\frac{q}{m}\right)\right]^n$$

where $q$ is the mass of substrate (M) adsorbed on the carbon at any time. An expression for $q$ is as follows:

$$q = q_o - V\left[(C - C_o) + \frac{\Delta G}{Y}\right]$$

where $q_o$ is the initial mass of substrate adsorbed on the carbon (M), $C_o$ is the initial concentration of the substrate in the bulk liquid (ML$^{-3}$), and $\Delta G$ is the biomass grown during regeneration (ML$^{-3}$). To determine $\Delta G$, MB on biomass can be written as follows:

$$\frac{dG}{dt} = \mu X$$

and

$$\frac{dX}{dt} = (\mu - b)X$$

where $b$ is the biomass death rate $(T^{-1})$. The regeneration efficiency, $R$ (%), can be determined from:

$$R = 100\left(\frac{q_o - q}{q_o}\right)$$

The model equations contain coupled algebraic equations and ODEs with nonlinear terms. They can be solved readily using dynamic simulation-based packages such as ithink®, Extend™, or Simulink®. In this example, the model developed with ithink® is illustrated in Figure 8.31. The three ODEs for $C$, $X$, and $G$ are represented by *Stocks* and are interconnected according to the equations. The fourth-order Runge-Kutta method is chosen to solve the model over a time span of 96 hours and a time step of 0.05 hours. The equations underlying the model and the model parameters are automatically compiled by ithink® into a list as shown in Table 8.4.

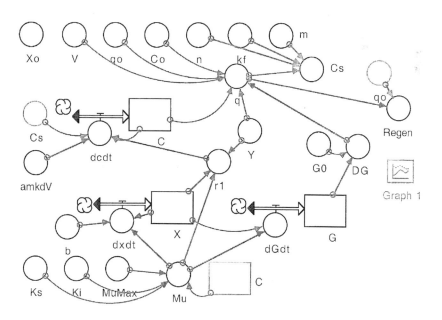

**Figure 8.31** Bioregeneration process modeled in ithink®.

Table 8.4  Bioregeneration Model Equations Generated by ithink®

```
C(t) = C(t – dt) + (Noname_1) * dt
     INIT        C = Co          INFLOWS:      dcdt = r1 + amkdV*(Cs–C)
G(t) = G(t – dt) + Noname_3) * dt
     INIT        G = G0          INFLOWS:      dGdt = Mu*X
X(t) = X(t – dt) + (Noname_2) * dt
     INIT        X = Xo          INFLOWS:      dxdt = (Mu–b)*X

Model Parameters:
     amkdV = 12.6;     b = .002      Co = 350;   V = 1                 qo = 4;
     G0 = 10;          kf = .0585;   n = 4.65;   Ki = 200;             Ks = 150;
     m = 20;           Xo = 200;     Y = .6      MuMax = .175;

     Regen = 100*(qo–q)/qo;                      r1 = –Mu*X/Y
     q = qo–(V*((C–Co)+ΔG/Y))/1000;              Cs = ((q/m)/kf)^n;
     Mu = MuMax*C/(Ks+C+(C^2)/Ki);               ΔG = G–G0
```

In this example, ithink®'s built-in feature for conducting sensitivity studies is demonstrated. By selecting any one of the model parameters at a time and assigning a range of values to it and selecting the *SensiRun* option, ithink® will run the model repeatedly for each of the values in the range and produce the results for comparison. For example, the sensitivity of the OBR process to initial biomass concentration, $X_o$, is examined in this case, and the results are shown in Figure 8.32. These model results are comparable to those presented by Goeddertz et al. (1988) who validated their model with experimentally measured data.

## 8.10 MODELING EXAMPLE: PIPE FLOW ANALYSIS

In this example, an analysis of sewer flow is illustrated. It is a common problem in environmental systems and has often been solved using graphical methods or tabulated data. Even though the governing equations are relatively simple and algebraic, their solution is cumbersome due to power terms and a trigonometric term. The following equations are well known in sewer flow analysis:

$$\text{Velocity} = \frac{1}{n}(\text{Hydraulic radius})^{\frac{2}{3}}(\text{slope})^{\frac{1}{2}}$$

$$\text{Hydraulic radius} = \frac{\text{Wetted perimeter}}{\text{Area}}$$

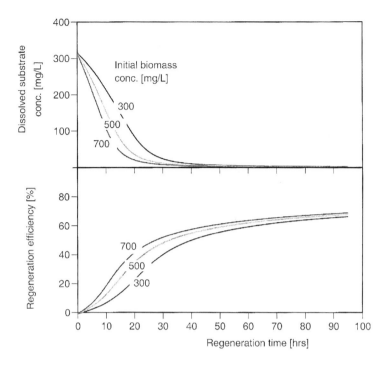

**Figure 8.32** Results of bioregeneration model from ithink®.

Because flow in sewers is often only partially full, the area, the wetted perimeter, and, hence, the hydraulic radius have to be determined from the depth of flow, *d,* and the pipe diameter, *D,* as follows:

$$\text{Area} = D^2\left[\left(\frac{\pi\varphi}{1440}\right) - \left(0.5 - \left(\frac{d}{D}\right)\sqrt{\left(d - \left(-\frac{d}{D}\right)^2\right)}\right)\right]$$

$$\text{Wetted perimeter} = \frac{\pi D\varphi}{360}$$

$$\text{Hydraulic radius} = \left(\frac{D}{4}\right) - \left(\frac{360D}{\pi\varphi}\right)\left(0.5 - \left(\frac{d}{D}\right)\right)\left[\sqrt{\left(\frac{d}{D}\right) - \left(\frac{d}{D}\right)^2}\right]$$

$$\text{where } \varphi = 2\cos^{-1}\left(1 - 2\left(\frac{d}{D}\right)\right)$$

Spreadsheet packages and equation solver-based packages can be readily used to solve the above equations. However, with spreadsheet packages, the equations are set up in a certain order and have to be solved in that order, with specific inputs. With the iterative solver option in Excel®, one variable at a time can be "backsolved." Thus, different models have to be set up in Excel® to evaluate various flow conditions. In the case of the TK Solver package, its unique feature of backsolving enables the same model to be used for a multitude of scenarios.

The model setup in TK Solver is illustrated in Figure 8.33. The equations listed above are entered first into the *Rules* sheet. Note that it is not necessary to follow any sequence among those equations. TK Solver automatically generates the list of variables in the *Variables* sheet. The model requires any four of the nine variables to be specified to determine the other five. In the special case where the depth of flow is unknown, a guess value has to be provided in order for TK Solver to complete the iterative process and establish the correct value for the depth and the other unknowns.

The advantage of the TK Solver model is that users can use this model, for example, to find velocity of flow at a given flow rate to check if self-cleansing is possible or to find the minimum slope to maintain a certain velocity or to find the diameter of the sewer to carry a certain flow at a certain velocity. These scenarios are evaluated by entering known parameters under the input column and running the model. If the model is unable to solve the equations

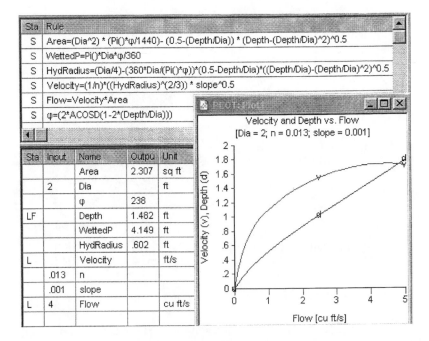

**Figure 8.33** Pipe flow analysis in TK Solver.

under a set of known data, the status column in the *Rules* sheet will indicate the rule that was not satisfied. In those cases, the status column in the *Variables* sheet may be marked with "G," and a guess value may be entered in the variables sheet for one of the variables, so that TK Solver can iteratively find the correct solution to satisfy all of the equations listed.

In the model shown, the flow is specified as a *List,* whereby the model can solve the same equations for a range of flow values that the user prefers. As shown, the depth and the velocity are also specified as *Lists,* so that they will be filled by TK Solver as each flow value specified in its list is used in the solution process. Then, the results stored in those *Lists* are used to plot graphs of velocity vs. flow and depth vs. flow as shown in Figure 8.33.

This basic model can be easily modified to generate the well-known hydraulic elements graph. By entering additional *Rules* for calculating the full flow parameters and the ratios for the hydraulic elements in the *Rules* sheet, and declaring those as *Lists,* in the *Variables* sheet, the model can be run to calculate the data for generating the graph. The modified model is shown in Figure 8.34, and the results from this model are used to generate the graph as shown in Figure 8.35.

**Rules**

| Sta | Rule |
|-----|------|
| S | Area=(Dia^2) * (Pi()*φ/1440)- (0.5-(Depth/Dia)) * (Depth-(Depth/Dia)^2)^0.5 |
| S | HydRadius=(Dia/4)-(360*Dia/(Pi()*φ))*(0.5-Depth/Dia)*((Depth/Dia)-(Depth/Dia)^2)^0.5 |
| S | WettedP=Pi()*Dia*φ/360 |
| S | Velocity=(1/n)*((HydRadius)^(2/3)) * slope^0.5 |
| S | Flow=Velocity*Area |
| S | φ=(2*ACOSD(1-2*(Depth/Dia))) |
| S | WettedPFull =Pi()*Dia |
| S | AreaFull =(Pi()*Dia^2)/4 |
| S | HydRadiusFull=AreaFull/WettedPFull |
| S | VelocityFull=(1/n)*(HydRadiusFull^(2/3)) * slope^0.5 |
| S | FlowFull = AreaFull*VelocityFull |
| S | WPRatio=WettedP/WettedPFull |
| S | ARatio=Area/AreaFull |
| S | HRRatio=HydRadius/HydRadiusFull |
| S | VRatio=Velocity/VelocityFull |
| S | FRatio=Flow/FlowFull |
| S | DRatio=Depth/Dia |

**Variables**

| Sta | Input | Name |
|-----|-------|------|
| L | | WettedP |
| L | | HydRadius |
| L | | Velocity |
| | .013 | n |
| | .001 | slope |
| L | | Flow |
| L | | WettedPFull |
| L | | AreaFull |
| L | | HydRadiusFull |
| L | | VelocityFull |
| L | | FlowFull |
| L | | WPRatio |
| L | | ARatio |
| L | | HRRatio |
| L | | VRatio |
| L | | FRatio |
| L | | DRatio |

**Figure 8.34** Pipe flow analysis in TK Solver.

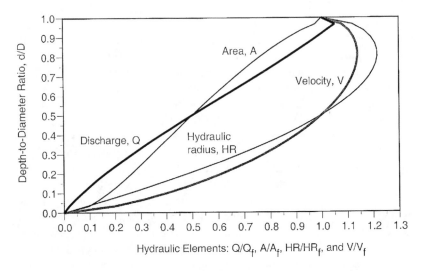

**Figure 8.35** Results of pipe flow analysis in TK Solver.

## 8.11 MODELING EXAMPLE: OXYGEN/NITROGEN TRANSFER IN PACKED COLUMNS

Oxygenation of aquacultural waters using commercial oxygen has been demonstrated to be beneficial, both economically and physiologically. Using commercial oxygen, the carrying capacity of hatcheries can be significantly increased at nominal costs, and harmful nitrogen can be removed simultaneously. One of the most effective process configurations is a countercurrent packed tower, in which the oxygen-deficient water can flow from the top of the tower, while commercial oxygen is blown from the bottom. In this example, a model for describing the concentration profiles of oxygen and nitrogen along the packing depth is developed for design and operation purposes.

The model is based on the Two-Film Theory, and it begins with elemental MBs on oxygen and nitrogen in the liquid phase (Nirmalakhandan et al., 1988; Speece et al., 1988):

MB across element on oxygen in liquid phase:

$$\frac{dC_o}{dz} = -\frac{(K_L a)_o}{u}(C_o^* - C_o)$$

MB across element on nitrogen in liquid phase:

$$\frac{dC_N}{dz} = -\frac{(K_L a)_N}{u}(C_N^* - C_n)$$

MB across element on oxygen in gas phase:

$$\frac{dm_o}{dz} = u\frac{dC_o}{dz}$$

MB across element on nitrogen in gas phase:

$$\frac{dm_N}{dz} = u\frac{dC_N}{dz}$$

where $C$ is the liquid phase concentration $(ML^{-3})$; $K_La$ is the mass transfer coefficient $(T^{-1})$; $m$ is the superficial gas flow rate; subscripts $O$ and $N$ represent oxygen and nitrogen, respectively; superscript * represents the liquid-phase concentration that is in equilibrium with the gas phase concentration; $u$ is the superficial liquid velocity $(LT^{-1})$; and $z$ is the packing height. Along the packing height, $z$, as the oxygen is transferred to the liquid phase, nitrogen is stripped from the liquid phase. Hence, the compositions of both the liquid and gas phases will change as a function of depth. Or, in other words, $C_O$, $C_N$, $m_O$, $m_N$, $C^*_O$, and $C^*_N$ will all be functions of $z$. Hence, additional equations are required to solve the above four coupled ODEs.

First, assuming that mass transfer coefficients are proportional to (diffusivity)$^{0.5}$ and that diffusivities are proportional to (molar volume)$^{0.6}$, a relationship between the mass transfer coefficients for oxygen and nitrogen can be derived as follows:

$$(K_La)_N = 0.942(K_La)_o$$

Then, introducing the gas phase mole fraction, $y$, and the air-water partition coefficient, $H$, the following relationships can be derived as follows for oxygen and nitrogen:

$$y_o = \frac{\dfrac{m_o}{32}}{\dfrac{m_o}{32}+\dfrac{m_N}{28}} = \frac{m_o}{m_o+\dfrac{8}{7}m_N}$$

$$C^*_o = \frac{y_o}{H_o} = \frac{1}{H_o}\left\{\frac{m_o}{m_o+\dfrac{8}{7}m_N}\right\}$$

$$y_N = \frac{\dfrac{m_N}{28}}{\dfrac{m_o}{32}+\dfrac{m_N}{28}} = \frac{m_N}{\dfrac{7}{8}m_o+m_N}$$

$$C^*_o = \frac{y_N}{H_N} = \frac{1}{H_N}\left\{\frac{m_N}{\dfrac{7}{8}m_o+m_N}\right\}$$

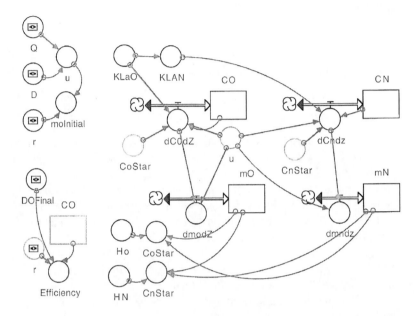

**Figure 8.36** Oxygen/nitrogen transfer modeled in ithink®.

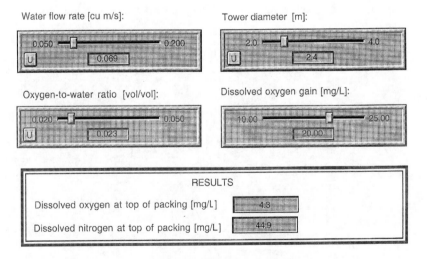

**Figure 8.37** User interface for oxygen/nitrogen model in ithink®.

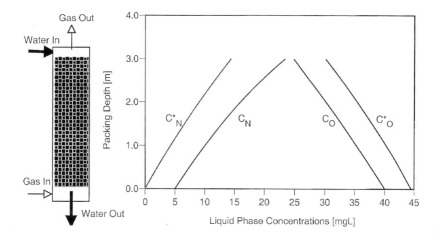

**Figure 8.38** Typical results of oxygen/nitrogen model in ithink®.

To design a tower of packing height, $z$, to achieve a desired gain in dissolved oxygen, the known variables are the $C$ values in the liquid phase at the top of the tower, $z = Z_{top}$, and $C_N = 0$ at the bottom of the tower, $z = z_{bottom} = 0$. The superficial flow rate of water to be oxygenated, $u$, and the gas-to-liquid ratio can be two parameters to be selected. The air-water partition coefficients can be found from handbooks; the mass transfer coefficients can be estimated from correlations. The above equations can be solved by an iterative trial-and-error process, by assuming the $C_N$ at the bottom of the packing, $z = 0$, and checking the $C_N$ value at the top. Computer implementation can greatly facilitate this iterative process.

In this example, the equations are implemented in the ithink® simulation package to take advantage of its interactive capabilities. The model diagram is shown in Figure 8.36, the user interface is shown in Figure 8.37, and typical results are shown in Figure 8.38.

## 8.12 MODELING EXAMPLE: GROUNDWATER FLOW MANAGEMENT

A common approach for groundwater cleanup is to extract the contaminated water and treat it for use or return the treated water by reinjection or release to surface drainage. After characterizing the contaminated site, several questions have to be answered before a system can be designed for this purpose. Some of the questions to be answered concern the optimum number and location of wells required, the optimum pumping rate, and the location for reinjection. Potential theory can be used as a first step in guiding us in this process. This example illustrates the application of potential and stream functions, introduced in Section 6.22 in Chapter 6, in visualizing flow patterns in groundwater flow management.

Consider a situation in which a contamination plume has been detected in an aquifer where a uniform flow exists in the x-direction. It is desired to locate pumping wells on the y-axis symmetrically about the x-axis. The goal here is to determine a combination of pumping rates and well spacing to ensure that no contamination will pass between the wells. This can be achieved analytically in the case of two wells as follows: by superposing the potential function for the uniform flow field with that for the two wells, the composite potential function can be obtained as follows:

$$\phi = -ux + \frac{Q}{4\pi} \ln\left[x^2 + \left(y - \frac{L}{2}\right)^2\right] + \frac{Q}{4\pi} \ln\left[x^2 + \left(y + \frac{L}{2}\right)^2\right]$$

where $Q$ is the pumping rate at each well, and $L$ is the spacing between the wells. The condition of no flow between the wells implies that there is a single stagnation point between the wells, or in terms of the potential and stream functions,

$$u(x,0) = 0$$

or

$$\left(\frac{\partial \phi}{\partial x}\right)_{x,0} = 0$$

The task of differentiating the expression for $\phi$, setting the resulting expression to zero, and then solving it for $x$ to check for stagnation conditions, although straightforward, can be tedious. However, Mathematica® can be used readily as shown in Figure 8.39 to differentiate the potential function, set the result to zero, and solve for $x$, to give the answer for $x$ to check for the location of the stagnation point. In line *In[1]* in Figure 8.39, the general expression for $\phi$ is entered, and Mathematica® is asked to substitute $y = 0$ in the expression (indicated by the symbol /. {y→ 0}), to find $\phi(x,0)$. The result is returned in line *Out[1]*. Then, in line *In[2]*, Mathematica® is asked to take the last result (indicated by the % symbol), differentiate it with respect to $x$ (indicated by the symbol $D$), set it to zero, and solve the result to find $x$. Mathematica® performs this sequence of operations in symbolic form and returns the result in line *Out[2]* as two possible roots for $x$. This example illustrates the ability of Mathematica® to present equations in two-dimensional form and to perform standard mathematical calculi in pure symbolic form.

To ensure only one root for the quadratic equation, the terms within the square root sign should cancel one another, or translating this mathematical

In[1]:= $-u \, x + qTerm \, Log\left[x^2 + \left(y - \frac{L}{2}\right)^2\right] +$

$qTerm \, Log\left[x^2 + \left(y + \frac{L}{2}\right)^2\right] \, /. \, \{y \to 0\}$

Out[1]= $-u \, x + 2 \, qTerm \, Log\left[\frac{L^2}{4} + x^2\right]$

In[2]:= $Solve[D[\%, x] == 0, x]$

Out[2]= $\left\{\left\{x \to \frac{16 \, qTerm - \sqrt{256 \, qTerm^2 - 16 \, L^2 \, u^2}}{8 \, u}\right\},\right.$

$\left.\left\{x \to \frac{16 \, qTerm + \sqrt{256 \, qTerm^2 - 16 \, L^2 \, u^2}}{8 \, u}\right\}\right\}$

**Figure 8.39** Mathematica® script for symbolic calculations.

interpretation to the real system, the spacing for the desired goal will be given by the following:

$$\sqrt{64 \, q\text{Term}^2 - 16 \, L^2 u^2} = 0$$

or

$$L = 2\left(\frac{q\text{Term}}{u}\right)$$

Alternatively, the entire analytical calculi can be readily conducted graphically by plotting the composite stream function:

$$\psi = -uy + \frac{Q}{2\pi} \tan^{-1}\left(\frac{y - \frac{L}{2}}{x}\right) + \frac{Q}{2\pi} \tan^{-1}\left(\frac{y + \frac{L}{2}}{x}\right)$$

The implementation of this approach using Mathematica® is illustrated in Figure 8.40. In this approach, numerical values for the variables are declared first, and the built-in *ContourPlot* routine is called as before to generate the streamlines. By adjusting the values of the variables, users can gain valuable insights through the visual interpretations of the above equations. This approach can be more effective and easier to apply than the analytical approach if more than two wells are to be evaluated.

This problem can be readily modeled with Mathcad® as well, as shown in Figure 8.41. The model parameters are first declared. A grid network is then set up in the *x*- and *y*-directions to plot the contours. The equation is now

```
In[66]:= u = 1.0;qTerm = 600.0; d=1000.0;
        ContourPlot[ψ[x,y] = qTerm *ArcTan[(y-d/2)/x]+
        qTerm*ArcTan[(y+d/2)/x]-u*y,
        {x,-2000.0,2000.0}, {y,-2000.0,2000.0},
        PlotPoints->100, Contours->20,ContourLines-
        >True,ContourShading->False, AspectRatio->0.6]
```

**Figure 8.40** Groundwater flow modeled in Mathematica®.

$u := 1$    $qTerm := 600$    $d := 1000$    $j := 0,1..50$    $k := 0,1..50$

$$x_j := -2000 + 80 \cdot j \qquad y_k := -2000 + 80 \cdot k$$

$$\psi(x,y) := qTerm \cdot atan\left[\frac{\left(y - \dfrac{d}{2}\right)}{x}\right] + qTerm \cdot atan\left[\frac{\left(y + \dfrac{d}{2}\right)}{x}\right] - u \cdot y$$

$$\psi\psi_{j,k} := \psi\left(x_j, y_k\right)$$

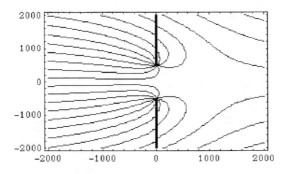

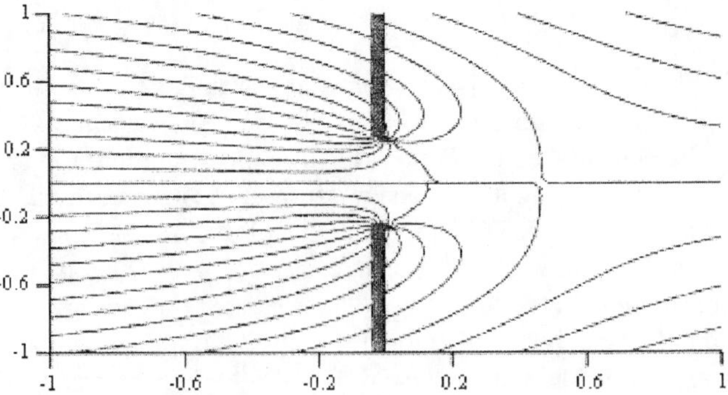

**Figure 8.41** Groundwater flow modeled in Mathcad®.

entered as a function of $x$ and $y$, using standard one-line notations, while Mathcad® automatically formats it to a two-dimensional form. The next line essentially calculates the value of $\psi$ at each of the node points. The plotting of the contours and the customizing are done through a GUI, without any scripting.

## 8.13 MODELING EXAMPLE: DIFFUSION THROUGH POROUS MEDIA

As a final example, the diffusion of vapors through porous media is modeled. Examples of such processes include emissions from landfills, drying out of land spills, etc. This phenomenon can be analyzed assuming one-dimensional diffusion of the vapors through the air-filled pores. The MB equation results in a partial differential equation:

$$\frac{\partial C}{\partial t} = \left(\frac{D_y}{\varepsilon}\right)\frac{\partial^2 C}{\partial y^2}$$

where $C$ is the gas phase concentration of the diffusing chemical (ML$^{-3}$), $D_y$ is the diffusion coefficient in the $y$-direction (L$^2$T), and $\varepsilon$ is the void fraction in the medium (–). The solution to the above PDE will require one initial condition (IC) and two boundary conditions (BC).

Consider a case in which a land spill has occurred and the pore spaces are instantly saturated with the chemical vapor. This scenario and its analysis have been described by Thibodeaux (1996). Suppose it is desired to develop the concentration profile in the soil as a function of time and to estimate flux at the soil-air interface and the time it would take for the concentration in the pore gases to drop to desired level, then, the appropriate IC and BCs for this condition can be as follows:

| | | |
|---|---|---|
| Initial condition: | $C = C_0$ | at $t = 0$ for all $y$ |
| Boundary condition (1): | $C = 0$ | at $y = h$ for all t |
| Boundary condition (2): | $\dfrac{\partial C}{\partial y} = 0$ | at $y = 0$ for all $t$ |

The IC is a result of the assumption that the void spaces are saturated initially. The first BC implies that the atmospheric concentration at the surface, $y = h$, is negligible at all times. The second BC stems from the assumption of an impervious boundary at $y = 0$, so that the flux at $y = 0$ is zero.

An analytical solution for this problem satisfying the above IC and BCs to describe the gas phase concentration as a function of depth and time has been reported, following an analogy with heat transfer (Thibodeaux, 1996):

$$C = C_o + (C_a - C_o)\frac{4}{\pi}\sum_0^\infty\frac{(-1)^n}{2n+1}\exp\left[-\frac{D_y(2n+1)^2\pi^2}{\epsilon h^2}t\right]\left\{\cos\left[\frac{(2n+1)\pi}{2h}\right]y\right\}$$

In the above form, the result is difficult to interpret or understand. However, it can be readily plotted in one of the equation solver packages such as Mathematica® as illustrated in Figure 8.42.

In this example, use of dynamic simulation packages in modeling PDEs is illustrated by modeling the above problem. (The examples presented so far involved single, coupled, first-order, and higher-order ODEs.) The dynamic simulation package, ithink®, is used here to model the above PDE. The ithink® package provides an elegant approach to model PDEs, wherein the *Stocks* are interconnected to represent the spatial variations, while the time variation within each *Stock* is modeled as usual. Increasing the number of *Stocks* can improve the accuracy of the solution. This approach is similar to a finite difference approach, except that the software does all the "account keeping."

```
In[1]:= diff = 0.005; pore = 0.6; h = 1; co = 0.1;
        c[y, t] =
          co +
          (0 - co) *
             10
             Σ   ( (-1)ⁿ      Exp[ -diff * (2*n+1)² * π² * t ]
            n=0  (2*n+1)              pore * h²

             Cos[ (2*n+1) * π * y ] ]);
                      2*h
ContourPlot[%, {t, 0, 20}, {y, 0, h}, PlotPoints → 50]
```

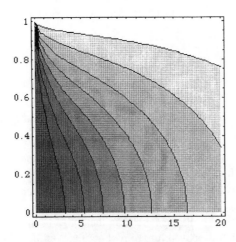

**Figure 8.42** Flow through porous media modeled in Mathematica®.

To generalize the approach, the governing PDE is first transformed into a nondimensional form. As alluded to in Chapter 1, by generalizing the governing equations, the same computer model can be easily adapted to solve other problems that have governing equations that can be reduced to the generalized form. In this example, this can be achieved by the following substitutions:

$$Y = \frac{y}{h}$$

and

$$T = \left(\frac{D_y}{\varepsilon}\right)\frac{t}{h^2}$$

resulting in the following nondimensional PDE:

$$\frac{\partial C}{\partial T} = r\frac{\partial^2 C}{\partial Y^2}$$

To implement the above equation in the finite difference form in ithink®, the last term of the above PDE has to be first expressed as follows:

$$\frac{C_{i+1,j} - 2C_{i,j} + C_{i-1,j}}{\Delta h^2}$$

By drawing the $C$ values at the $j$th time step from the consecutive *Stocks*, $i - 1$, $i$, and $i + 1$, the general equation can be easily implemented in the ithink® flow diagram as illustrated in Figure 8.43. Here, six *Stocks* are used in series to represent the soil layers. If necessary, different soil properties may be assigned to each layer.

The results from the ithink® model are plotted in Figure 8.44 (using the Deltagraph® plotting package) showing the concentration distribution within the layers as a function of time. A close resemblance can be noted between the plot generated by Mathematica®, analytically, and the plot generated by ithink®, numerically.

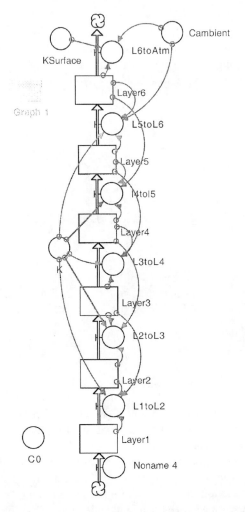

**Figure 8.43** Flow through porous media modeled in ithink®.

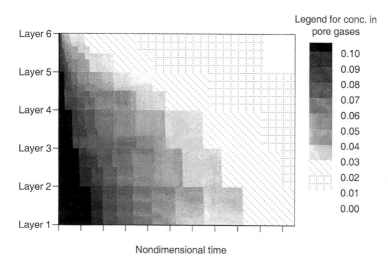

**Figure 8.44** Concentration in pore gases in the layers as a function of time.

## EXERCISE PROBLEMS

8.1. Refer to Exercise Problem 5.1. Solve the coupled differential equations derived in that problem using the following biokinetic data: $\mu_{max} = 0.3$ $hr^{-1}$, $k_d = k_r = 0.01$ $hr^{-1}$, $Y = 0.45$ gr cells/gr $C$ in substrate, and $K_s = 150$ mg/L. Hence, plot the temporal variation of biomass and substrate.

Assume that the initial concentrations of biomass and substrate are 5 mg/L as $C$ and 1000 mg/L as $C$, respectively.

8.2. Refer to Exercise Problem 5.2. Solve the coupled differential equations derived in that problem using the same biokinetic data in Exercise Problem 8.1 for three different hydraulic retention times (HRT) of 5 hr, 10 hr, and 20 hr. Hence, plot the temporal variation of biomass and substrate for each HRT.

Use the following additional data: $V = 10$ L; $S_{in} = 1000$ mg/L as $C$; and initial concentrations of biomass and substrate in the reactor are 5 mg/L and 0 mg/L as $C$, respectively.

## CHAPTER 9

# Modeling of Natural Environmental Systems

## CHAPTER PREVIEW

*In this chapter, 12 examples of natural systems are illustrated. The selected examples include steady and unsteady state analysis using algebraic and differential equations, solved by analytical, trial-and-error, and numerical methods. Computer implementation of the mathematical models for the above are presented. The rationale for selecting appropriate software packages for modeling the different problems and their merits and demerits are discussed.*

## 9.1 INTRODUCTION

In this chapter, the modeling of several examples of natural environmental systems using the three types of software packages are demonstrated. The use of Excel®, TK Solver, Mathcad®, Mathematica®, MATLAB®, Extend™, ithink®, and Simulink® software packages in modeling aquatic, soil, and atmospheric systems under various conditions are illustrated. The development of the mathematical models in each case is outlined, and the rationale for the selection of the software packages for each example, their ease of use, applicability, and limitations are pointed out. The examples included here demonstrate how these software packages can be used to solve various mathematical calculi commonly encountered in environmental modeling.

## 9.2 MODELING EXAMPLE: LAKES IN SERIES

The basic lake models discussed in Chapter 7 can be easily modified and refined in stages to simulate more complex and realistic situations. The modeling of a sample problem from Thomann and Mueller (1987) involving two lakes in series is illustrated in this example. A constant load of a conservative substance ($K = 0$) had been applied to the first lake resulting in concentrations of 0.270 mg/L in that lake and 0.047 mg/L in the second lake. Then, the load to the first lake is instantaneously removed. The goal is to develop a model to describe the temporal changes in the concentrations in the two lakes.

The MB equations for the two lakes now yield the following coupled differential equations:

$$\frac{dC_1}{dt} = \frac{W_{1(t)}}{V_1} - \left(\frac{Q_{1,2}C_1 + V_1 K_1 C_1}{V_1}\right) = \frac{W_{1(t)}}{V_1} - \alpha_1 C_1$$

$$\frac{dC_2}{dt} = \frac{W_{2(t)}}{V_2} + \frac{Q_{1,2}C_1}{V_2} - \left(\frac{Q_2 C_2 + V_2 K_2 C_2}{V_2}\right) = \frac{W_{2(t)}}{V_2} + \frac{Q_{1,2}C_1}{V_2} - \alpha_2 C_2$$

where $C$ is the concentration in the lake ($ML^{-3}$), $V$ is the volume ($L^3$), $K$ is the overall first-order reaction rate constant ($T^{-1}$), subscripts 1 and 2 represent the first and second lake, $Q_{1,2}$ is the flow rate from lake 1 to lake 2, and $Q_2$ is the flow rate from lake 2. These coupled ODEs can be analytically solved for certain simple input functions as illustrated by Thomann and Mueller (1987). In the current example, they can be solved for the concentration in the second lake due to the washout of the first lake, $C_{2,1}$, and the concentration due to its own washout, $C_{2,2}$. The following result has been reported by Thomann and Mueller (1987):

$$C_2 = \{C_{2,1}\} + \{C_{2,2}\} = \left\{\alpha_{1,2}C_{1,0}\left[\frac{e^{-\alpha_1 t}}{(\alpha_2 - \alpha_1)} + \frac{e^{-\alpha_2 t}}{(\alpha_1 - \alpha_2)}\right]\right\} + \{C_{2,0}[e^{-\alpha_2 t}]\}$$

The system response is not obvious from the above equation and is counterintuitive: the concentration in the second lake increases for a short period in response to the shutdown. Spreadsheet packages can be readily used to model this particular case, because the analytical solution to the governing ODEs is known for the simple step shutdown of the input function to the first lake. To model other scenarios, a numerical solution procedure may have to be used. Implementing a numerical procedure such as the Runge-Kutta method in a spreadsheet for this problem may be tedious. Equation solver-type packages and dynamic simulation packages are more suitable for modeling this problem under such conditions.

The use of equation solver-based packages would be possible only when the input parameters are not arbitrary functions of time. As a first example, the model developed with Mathematica® is shown in Figure 9.1. In this case, Mathematica®'s built-in function, *DSolve*, is used to find the analytical solution for the two coupled ODEs as shown in line *In[1]*. The result returned in line *Out[5]* can be seen to be identical to the above analytical solution. This

In[1]:= **HRT1 = 99.1; HRT2 = 22.68; Q12 = 51900; Q2 = 175400;**
**K1 = 0.; K2 = 0.;**
**α1 = K1 + 1/HRT1; α2 = K2 + 1/HRT2; αα12 =**
**Q12/(HRT2*Q2);**
**w1 = 0.; w2 = 0.; V1 = 1.72; V2 = 1.25;**
**DSolve[ {Derivative[1][c1][t] == w1/V1 - α1*c1[t],**
**Derivative[1][c2][t] == w2/V2 + αα12*c1[t] -**
**α2*c2[t],**
**c1[0] == 0.27, c2[0] == 0.047}, {c1[t], c2[t]}, t]**

Out[5]= $\{\{c1[t] \to 0.27\, e^{-0.0100908\, t},\ c2[t] \to 1.48918 \times 10^{-19}$
$e^{-0.0440917\, t}\ (-3.80088 \times 10^{17} + 6.95698 \times 10^{17}\ e^{0.0340009\, t})\}\}$

In[6]:= **Plot[Evaluate[{c1[t], c2[t]} /. %], {t, 0, 200},**
**AxesLabel → {"Time [yrs]", "Conc.in lakes [mg/L]"},**
**PlotRange → All, AxesOrigin → {0, 0}]**

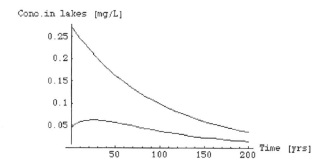

**Figure 9.1** Two lakes in series modeled in Mathematica®.

example illustrates the unique and powerful feature of Mathematica® in solving coupled ODEs, analytically, in symbolic form. A plot of concentration vs. time for the two lakes is generated using the commands in line *In[6]*.

The Mathcad® model for the above scenario is shown in Figure 9.2. Because Mathcad® cannot solve the coupled ODEs analytically, a built-in numerical routine, *rkfixed,* which is based on the Runge-Kutta method, has to be used. The right-hand sides of the governing ODEs are specified as a vector, *D,* in the call to *rkfixed,* which returns the solution as a vector *C.*

MATLAB® is also unable to find the analytical solution. Hence, a numerical approach is used as shown in Figure 9.3. Here, an M-File is first created in which the model parameters are declared in lines 2 to 6. Line 7 contains the right-hand side of the two ODEs to be solved. The built-in numerical procedure, *ode45,* is called from the *Command* window, with the following arguments: the name of the M-File containing the model parameters and the equations, the range of the independent variable over which the solution is

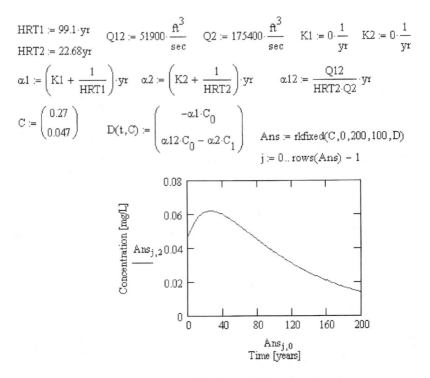

$$HRT1 := 99.1 \cdot yr \qquad Q12 := 51900 \cdot \frac{ft^3}{sec} \qquad Q2 := 175400 \cdot \frac{ft^3}{sec} \qquad K1 := 0 \cdot \frac{1}{yr} \qquad K2 := 0 \cdot \frac{1}{yr}$$

$$HRT2 := 22.68yr$$

$$\alpha 1 := \left( K1 + \frac{1}{HRT1} \right) \cdot yr \qquad \alpha 2 := \left( K2 + \frac{1}{HRT2} \right) \cdot yr \qquad \alpha 12 := \frac{Q12}{HRT2 \cdot Q2} \cdot yr$$

$$C := \begin{pmatrix} 0.27 \\ 0.047 \end{pmatrix} \qquad D(t,C) := \begin{pmatrix} -\alpha 1 \cdot C_0 \\ \alpha 12 \cdot C_0 - \alpha 2 \cdot C_1 \end{pmatrix} \qquad Ans := rkfixed(C,0,200,100,D)$$

$$j := 0 .. rows(Ans) - 1$$

**Figure 9.2** Lakes in series modeled in Mathcad®.

sought, and the initial values for the two equations. The subsequent commands generate a plot of the results returned by the call to *ode45*.

If different loading conditions are to be evaluated, dynamic simulation programs would be more appropriate for this problem. In this example, the lake model developed using the Extend™ dynamic simulation package in Chapter 7 is modified for the two lakes as shown in Figure 9.4.

This model allows a wide range of input functions to be specified through the *Function Input* blocks. Two sets of *Integration* blocks are used to solve each of the differential equations, the output from the first one acting as the input to the second one, in addition to its own external input. In the example shown in Figure 9.4, a step shutdown is specified for the first lake.

This model can be further expanded and refined to simulate more realistic situations. For example, the classical problem of lakes in series (e.g., the Great Lakes) could be set up by copying and duplicating the basic "lake block" already developed, and assigning individual parameters. Catchment areas may be added to estimate the inflows to the lakes due to runoff, with user-specified runoff characteristics and annual rainfall information

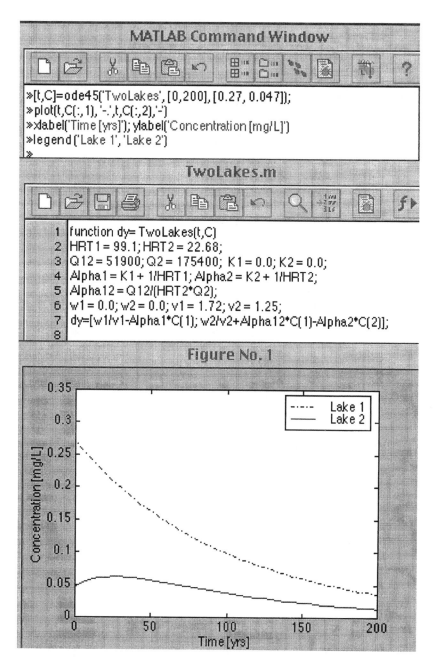

**Figure 9.3** Two lakes in series modeled in MATLAB®.

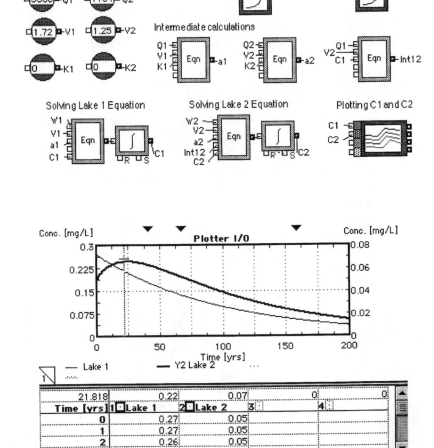

**Figure 9.4** Two lakes in series modeled in Extend™.

downloaded from a database via the *File Input* icon. Additional waste loads with random time variations can be readily added to the lakes. Submodels may be added to predict the impact on fish in the lake, buildup of sediment concentrations, etc.

To simplify the appearance of models with several icons, Extend™ allows related icons to be grouped and placed inside custom-designed icons as shown in Figure 9.5 for part of the Great Lakes system. Double clicking the drainage basin for Lake Superior reveals *Constant Input* icons for inputting

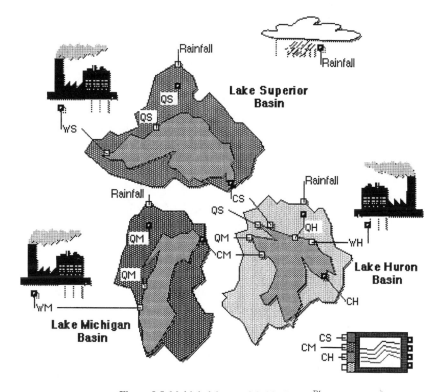

**Figure 9.5** Multiple lakes modeled in Extend™.

the runoff characteristics and an *Equation* icon where the equation for calculating the runoff has been entered by the developer. The output from the *Equation* icon, the runoff, is connected to the other icons that use that variable. This feature of customized graphic icons that encode the equations can provide strong visual appreciation and a global view of the problem.

## 9.3 MODELING EXAMPLE: RADIONUCLIDES IN LAKE SEDIMENTS

Radionuclides or radioactive substances have been released into the environment by anthropogenic activities such as energy generation, weapons development, and some industrial applications. They behave similar to organic chemicals except in the following regards: they do not volatilize readily, they undergo a decay process often by first order, and they are measured in curie units instead of mass.

In this example, a two-compartment model is developed to evaluate the impact of fallout of radionuclides resulting from nuclear weapons testing conducted in the late 1950s and early 1960s. The system is Lake Michigan and the sediments. The objective is to predict the long-term fate of cesium in the water column and the sediments. This illustration follows the mathematical model reported by Chapra (1997), which is based on the simplified system illustrated in Figure 9.6.

The MB equations for dissolved concentrations in the water column and the pore waters, and for solids in the water column and the sediments are as follows:

MB on cesium in dissolved form in water column:

$$\frac{dC_1}{dt} = \left(\frac{1}{V_1}\right)\{W_c - QC_1 - kC_1V_1 - v_sAf_{p,1}C_1 + EA(f_{d,2}C_2 - f_{d,1}C_1)\}$$

MB on cesium in dissolved form in pore waters:

$$\frac{dC_2}{dt} = \left(\frac{1}{V_2}\right)\{-kC_2V_2 + v_sAf_{p,1}C_1 - EA(f_{d,2}C_2 - f_{d,1}C_1) - v_rAC_2 - v_bAC_2\}$$

MB on particulates in water column:

$$\frac{dm_1}{dt} = \left(\frac{1}{V_1}\right)\{W_s - Qm_1 - v_sAm_1 + v_rAm_2\}$$

MB on particulates in sediments:

$$\frac{dm_2}{dt} = \left(\frac{1}{V_2}\right)\{v_sAm_1 - v_rAm_2 - v_bAm_2\}$$

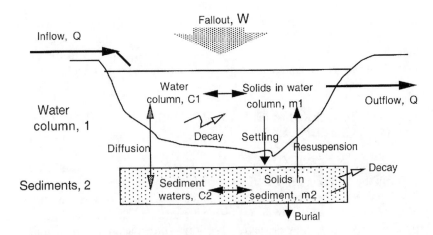

**Figure 9.6** Schematic diagram of lake-sediment system.

where $C_1$ and $C_2$ are the concentrations of cesium in the water column and the sediment waters, $V_1$ and $V_2$ are the volumes of water column and sediment, $W_c$ and $W_s$ are the input rates of cesium and solids, $Q$ is the outflow rate of water, $k$ is the first-order decay rate constant, $v_s$ and $v_r$ are the settling and resuspension velocities of solids, $A$ is the water-sediment interfacial area, $f_{p,1}$ and $f_{d,1}$ are the particulate and dissolved fractions in the water column, $E$ is the sediment-water column diffusion coefficient, $f_{d,2}$ is the fraction dissolved

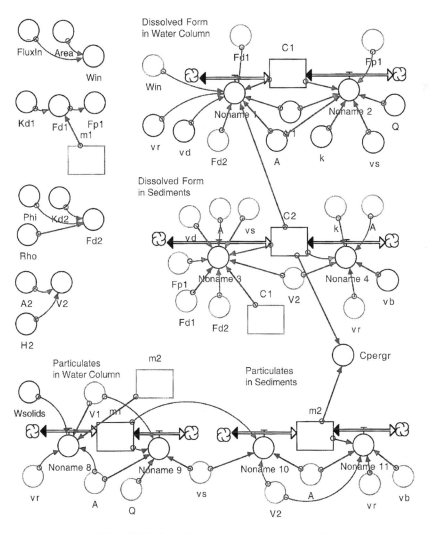

**Figure 9.7** Lake-sediment system modeled in ithink®.

in the sediments, $v_b$ is the burial rate constant, and $m_1$ and $m_2$ are the solids concentrations in the water column and the sediments.

The above first-order coupled differential equations can be solved numerically, using the Runge-Kutta method, for instance. The Excel® spreadsheet package or the equation solver-based packages can be used if the model coefficients (parameters) are constant and the forcing function, $W$, is a constant or a simple function of time. In this example, because the forcing function $W$ is an arbitrary function of time, dynamic simulation packages wold be most

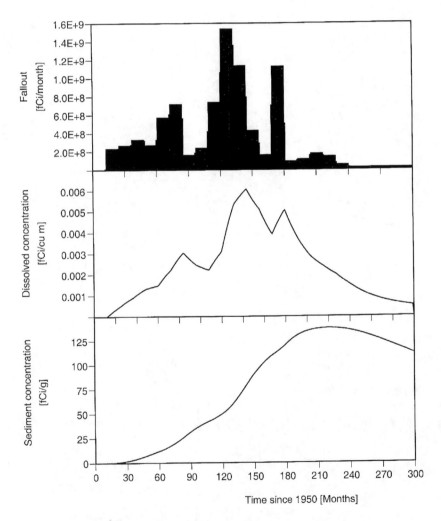

**Figure 9.8** Fallout, dissolved concentration, and sediment concentration of cesium.

suitable. The use of the ithink® software package is chosen in this example as illustrated in Figure 9.7.

The results from this model are presented in Figure 9.8. The slow buildup of the sediment concentration can be seen from this plot. These results are in agreement with those presented by Chapra (1997).

## 9.4 MODELING EXAMPLE: ALGAL GROWTH IN LAKES

In this example, development of a model to describe algal growth in lakes is illustrated, starting from a simple two-component model and gradually refining it to make it more realistic. The model assumes the constant volume of lake, $V(L^3)$; complete mixing; nutrient limiting conditions for algae growth; and phosphorous recycling to the phosphorous pool after death. The preliminary model assumes a constant maximum growth rate, $k_{g,max}$ $(T^{-1})$, and first-order death process of rate constant, $k_d$ $(T^{-1})$. The MB equations for algae, $a$, and phosphorous, $p$, are as follows:

$$\frac{da}{dt} = \left(k_{g,max}\frac{p}{K_{s,p} + p}\right)a - k_d a - \left(\frac{Q}{V}\right)a$$

$$\frac{dp}{dt} = a_{pa}(k_d a) + \left(\frac{Q}{V}\right)(p_{in} - p) - a_{pa}\left(k_{g,max}\frac{p}{K_{s,p} + p}\right)a$$

where $Q$ is the inflow rate $(L^3 T^{-1})$, $p_{in}$ is the influent phosphorous concentration $(ML^{-3})$, $K_{sp}$ is the half saturation constant for $p$ $(ML^{-3})$, and $a_{pa}$ is the stoichiometric ratio of $p$ in $a$ $(-)$. The term $(V/Q)$ can be replaced in terms of the hydraulic detention time, HRT = $V/Q$ (T). The above equations are coupled nonlinear differential equations and have to be solved numerically. Equation solver packages such as TK Solver, Mathcad®, MATLAB®, or Mathematica® can be used for solving them; however, if the model parameters such as $p_{in}$ are functions of time, dynamic simulation packages such as ithink® or Extend™ would be more efficient in solving them. In this example, the above equations are implemented in the ithink® dynamic simulation package.

The model developed using the ithink® simulation package is illustrated in Figure 9.9. The construction of the flow diagram is relatively straightforward. The model parameters are first assigned numerical values using the *Containers*. Two *Stocks* are created to represent a and p; the equations are entered into the *Converters,* which are automatically compiled by ithink® and are listed in Table 9.1. The flow diagram visually illustrates the interactions between the algae and phosphorous compartments, and at the same time, it encodes the underlying equations governing the system. The model is set to solve the differential equations using the Runge-Kutta fourth-order method for

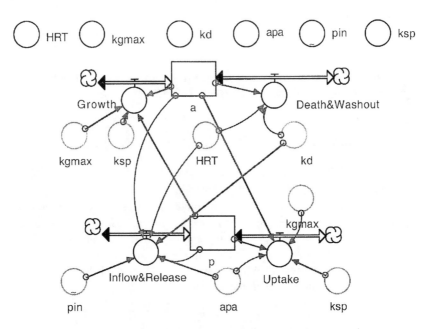

**Figure 9.9** Algal growth modeled in ithink®.

30 days, with a time step of 0.1 days. Temporal variations of concentrations of algae and phosphorous, resulting from a step input of $p_{in}$, are shown in Figure 9.10.

The Extend™ simulation package can also be easily used to model this problem as illustrated in Figure 9.11. Two *Integrator* blocks are used to solve the coupled differential equations: one for $a$ and the other for $p$. In this case, the influent concentration of phosphorous, pin, is set as a constant value of

Table 9.1 **Model Equations Compiled by ithink®**

```
A(t) = a(t - dt) + (Growth - Death&Washout) * dt
    INIT                  a = 0.5
    INFLOWS:  Growth = kgmax*(p/(p+ksp))*a
    OUTFLOWS:     Death&Washout = kd*a + (1/HRT)*a
p(t) = p(t - dt) + (Inflow&Release - Upake) * dt
    INIT                  p = 1
    INFLOWS:  Inflow&Release = apa*kd*a + (1/HRT)*(pin-p)
    OUTFLOWS:     Uptake = apa*kgmax* (p/(p+ksp))*a
apa = 1.5; HRT = 30; kd = 0.1; kgmax = 1; ksp = 2
```

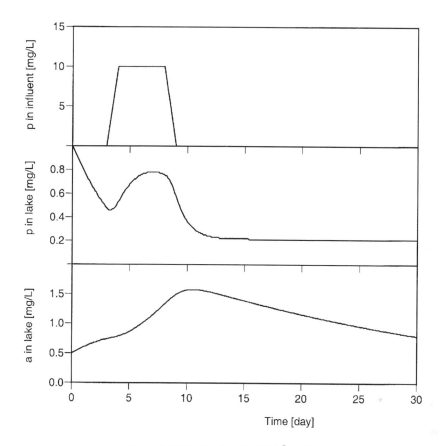

**Figure 9.10** Results from the ithink® model.

10 mg/m³. However, it can be readily set to be a variable as in the ithink® example, using a *Table* or *Function* input block.

Unlike in ithink®, the underlying equations are not compiled into a list in Extend™. The equations underlying each block can be viewed or edited by double clicking that block. Even though the general setup is nearly the same in ithink® and Extend™, the ithink® model is visually somewhat more compact than the Extend™ model.

The built-in *Plotter* block is used to generate a plot of the temporal variations of algae and phosphorous as shown in Figure 9.12. Unlike ithink®, an Extend™ plot has more features for reading the plot and customizing it. For example, by placing the cursor at any point within a graph, the coordinates of the points on the graph are displayed in the first row of the table below. The toolbar at the top of the plot allows access to several customizing features. In addition, the table associated with the plot is readily available for inspection.

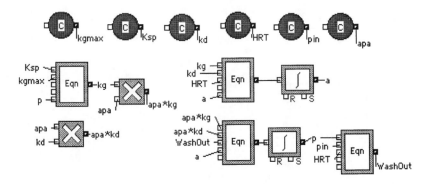

**Figure 9.11** Algal growth modeled in Extend™.

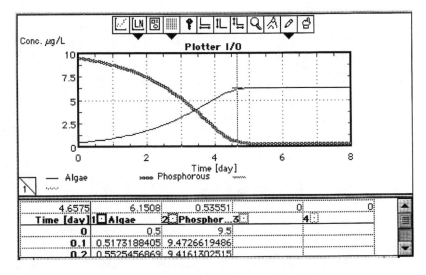

**Figure 9.12** Results of algal growth modeled in Extend™.

As the next step, a basic algae-zooplankton grazing model can be represented by the following two MB equations, assuming constant coefficients:

$$\frac{da}{dt} = (k_g - k_{ra})a - C_{gz}za$$

$$\frac{dz}{dt} = a_{ca}(C_{gz}za) - k_{dz}z$$

These two equations can be easily implemented in the ithink® simulation package, as shown in Figure 9.13. Temporal variations of phytoplankton-C, zooplankton-C, and total *C* predicted by the ithink® model are also included in Figure 9.13.

These examples demonstrate the ease with which dynamic models can be readily assembled for systems that would normally require extensive programming expertise for computer implementation by traditional languages.

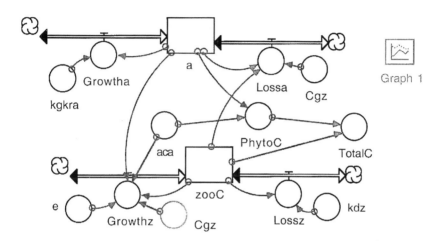

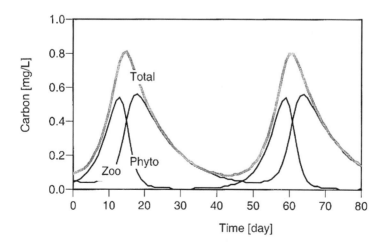

**Figure 9.13** Algae-zooplankton interaction modeled in ithink®.

Once the basic model is constructed, it can be refined in stages by increasing the complexity. For example, submodels describing the effect of temperature on the maximum growth rate, the influence of sunlight on algae growth rate, growth-controlling nutrient (nitrogen vs. phosphorous), and predation can be integrated to generate a comprehensive and more realistic model. The combined effect of temperature and sunlight on the growth rate of algae, for instance, can be modeled as follows:

$$k_G = k_{G,20} 1.066^{(T-20)} \left[ \frac{2.718f}{k_e H}(e^{\alpha_1} - e^{\alpha_0}) \right]$$

$$\alpha_0 = \left( \frac{I_a}{I_s} \right) e^{-k_e H_1}$$

and

$$\alpha_1 = \left( \frac{I_a}{I_s} \right) e^{-k_e H_2}$$

where $k_{G,20}$ is the growth rate measured at 20°C, $T$ is the temperature (°C), $k_e$ is the light extinction coefficient ($L^{-1}$), $H = H_2 - H_1$ is the depth of algae activity (L), and ($I_a/I_s$) is the ratio of average light level to the optimal light level (–).

The growth-controlling phenomenon of multiple nutrients can be incorporated by modifying the above expression for $k_G$ as follows:

$$k_G = k_{G,20} 1.066^{(T-20)} \left[ \frac{2.718f}{k_e H}(e^{\alpha_1} - e^{\alpha_0}) \right] \left\{ \text{minimum of} \left( \frac{n}{K_{s,n} + n}, \frac{p}{K_{s,p} + p} \right) \right\}$$

where $n$ is the concentration of available nitrogen ($ML^{-3}$), and $K_{s,n}$ is the half saturation constant for nitrogen ($ML^{-3}$). To use the above expression, a submodel for the fate of available nitrogen has to be included. The interaction between algal-carbon, $ca$, and available nitrogen, $n$, has been modeled by the following equations (Chapra, 1997):

$$\frac{dca}{dt} = 0.389 \left( 1.94 \frac{I_{av}}{53.5 + I_{av}} \frac{n}{K_{s,n} + n} \right) ca - 0.042ca - \left( \frac{Q}{V} \right) ca$$

$$\frac{dn}{dt} = -N/C \left[ 0.389 \left( 1.94 \frac{I_{av}}{53.5 + I_{av}} \frac{n}{K_{s,n} + n} \right) ca - 0.042ca \right] + \left( \frac{Q}{V} \right)(n_{in} - n)$$

$$I_{ave} = \frac{I_o}{k_e H}(1 - e^{-k_e H})$$

where $N/C$ is the nitrogen-carbon conversion ratio, and $n_{in}$ is the available nitrogen concentration in the influent ($ML^{-3}$).

As a next step, grazing of algae by zooplankton can be incorporated into the model by first developing a submodel assuming grazing loss as a first-order process with respect to algae concentration, where the rate constant is a function of zooplankton concentration, with a temperature correction factor:

$$\text{Grazing loss} = -k_{gz}za = -\left[\left(C_{gz}\frac{a}{K_{s,a}+a}\right)\theta_{gz}^{(T-20)}\right]za$$

where $K_{s,a}$ is the half saturation constant for the zooplankton grazing on algae ($ML^{-3}$). In order to use the above, an MB equation for zooplankton is required, assuming growth of zooplankton due to assimilation of algae and loss due to respiration and death. Thus, the MB equation based on zooplankton concentration, $z$, can be formulated as follows:

$$\frac{dz}{dt} = a_{ca}\varepsilon\left\{\left[\left(C_{gz}\frac{a}{K_{s,a}+a}\right)\theta_{gz}^{(T-20)}\right]za\right\} - k_{dz}z$$

where $a_{ca}$ is the ratio of carbon to chlorophyll $a$ in algae (–), $\varepsilon$ is the grazing efficiency factor, and $k_{dz}$ is the first-order rate constant for respiration and death ($T^{-1}$).

A more complete representation of the system is now possible with the above equations. With some practice, a comprehensive model could be generated with simulation packages such as ithink®, Extend™, or Simulink®. A model that incorporates algae, herbivorous zooplankton, carnivorous zooplankton, particulate organic carbon, dissolved organic carbon, ammonium-nitrogen, nitrate-nitrogen, and soluble phosphorous has been developed based on the research by Chapra (1997). This model is based on a total of eight coupled differential equations derived from MB on the above species in a lake, interacting as shown in Figure 9.14, and driven by seasonal variations in temperature and sunlight.

The graphical interface of this model developed with the ithink® package is illustrated in Figure 9.15. Results from a typical run (included in Figure 9.15) follow the general trend reported by Chapra (1997). The model requires several simplifying assumptions and over 30 input parameters (Chapra,

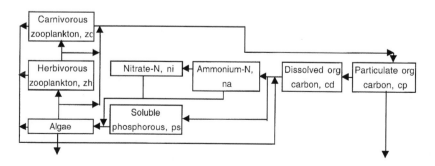

**Figure 9.14** Interactions between model compartments.

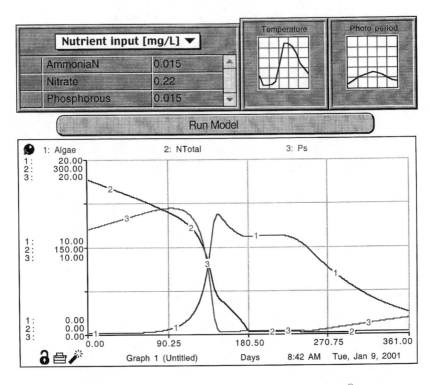

**Figure 9.15** Graphical user interface for a lake model in ithink®.

1997). A model such as this can be used to study the sensitivity of the various parameters for optimal design of experiments, to use in field studies, and to determine the impact of alternative management actions.

## 9.5 MODELING EXAMPLE: CONTAMINANT TRANSPORT VISUALIZATION

In this example, the use of software packages in visualizing a groundwater contamination site is presented. The contamination is caused by an accidental release of a mass $M$ of a chemical through a reinjection well. A mathematical model is to be developed to describe the fate of the contaminant in the aquifer. The model is expected to include advective transport as well as longitudinal and transverse dispersion and retardation and to be able to predict contaminant concentration as a function of space and time. The ultimate objective of the modeling exercise is to develop an appropriate model for use in visualizing the temporal and spatial distribution of the contaminant, to aid in risk assessment.

In this case, it will be assumed that the medium is uniform, advective flow is one-dimensional in the $x$-direction, the release is uniformly distributed vertically through the thickness of the aquifer (i.e., fully penetrating screened reinjection well), the spill volume is small compared to the aquifer volume, and occurs over a very short time. A material balance under these conditions simplifies to the following:

$$\frac{\partial C}{\partial t} + v\frac{\partial C}{\partial x} + \lambda C = D_{xx}\frac{\partial^2 C}{\partial x^2} + D_{yy}\frac{\partial^2 C}{\partial y^2}$$

The initial condition necessary to solve the above partial differential equation (PDE) can be found by a material balance at the point of discharge. While the above equation cannot be solved analytically by any of the software packages directly, an analytical solution reported by Charbeneau (2000) can be used for visualization:

$$C_{(x,y,t)} = \frac{V_0 C_0 \exp\left\{-\left[\frac{(x-vt)^2}{4D_{xx}t}\right] - \frac{y^2}{4D_{yy}t}\right\}}{4\pi bt\sqrt{D_{xx}D_{yy}}}\exp(-\lambda t)$$

The above result is not a straightforward one able to be understood intuitively. However, it can be visualized with most software packages discussed in this book. The equation solver-based packages Mathematica®, Mathcad®, and MATLAB®, with their powerful graphing capabilities, can be particularly efficient in visualizing the above result.

The application of Mathematica® to generate a variety of images to visualize this problem is summarized first. The basic Mathematica® script encoding the above equation is shown in Figure 9.16, where the variable, time, is set at 10.

Once the basic script is written to get the solution $C[x,y]$, the result can be used with a range of the Mathematica® built-in plot routines to develop

```
mass=50;vel=1.1; dxx=10; dyy =
1.0;n=0.3;b=5.;lambda=0.002;time=10;

c[x,y]={mass*1000* Exp[-((x-
vel*time)^2/(4*dxx*time*30.))-(y^2/(4*dyy*time*30.))
* Exp[-lambda*time*30.]) /
(4*Pi*n*b*time*30.*(dxx*dyy)^.5);
```

**Figure 9.16** Basic script in Mathematica® for contaminant transport.

```
Plot3D[X, {x,5,250}, {y,-100,100}, PlotRange->{0,3},
ViewPoint->{0,-2,0}]
```

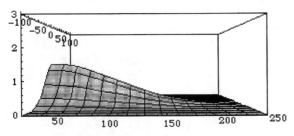

- SurfaceGraphics -

```
Show[ContourGraphics[X], ContourShading-> False]
```

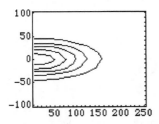

**Figure 9.17** Contaminant transport visualization in Mathematica®.

two- and three-dimensional graphs and animations to visualize the results for better understanding. For example, *Plot3D* routine and the *ContourGraphics* routine are illustrated in Figure 9.17.

The same basic code is slightly modified to generate a series of three-dimensional surfaces at increasing time steps ($t = 2$ to $10$) as shown in Figure 9.18. This series of figures can also be compiled to generate a movie to animate the spread of the plume.

The Mathcad® implementation of this example is shown in Figure 9.19. Again, the true equation-based interface in Mathcad® makes the model appear to be organized and easily readable. Once the governing equation is entered, the powerful graphing capabilities of Mathcad® with script-free formatting of the plots through a GUI makes Mathcad® most appealing for this example. The generation of the composite plot of a three-dimensional surface of the concentration distribution and the corresponding two-dimensional contours; the lighting, colors, and transparency of the surface; the thickness of the contour lines; the position of axes; and the viewing angle, are all accomplished through the GUI without any scripting.

```
Table[Plot3D[(mass*1000* Exp[-((x-
vel*time)^2/(4*dxx*time*30.))-
(y^2/(4*dyy*time*30.))] * Exp[-lambda*time*30.])
/ (4*Pi*n*b*time*30.*(dxx*dyy)^.5), {x,5,250},
{y,-100,100}, Axes-> False,PlotRange-
>{0,3},DisplayFunction -> Identity], {time,2,10}]

Show[GraphicsArray[Partition[%,3]]]
```

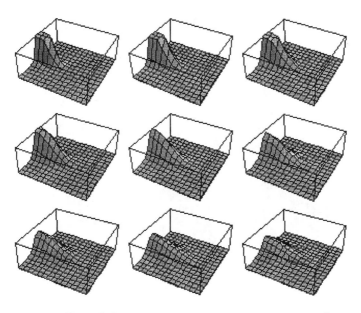

**Figure 9.18** Script for generating animation in Mathematica®.

## 9.6 MODELING EXAMPLE: METHANE EMISSIONS FROM RICE FIELDS

In this example, development of a model to predict methane release from rice fields is illustrated, which in turn, can be used in modeling greenhouse effects. (Methane emission is a serious environmental issue because it is 20 times more absorptive than carbon dioxide, and global methane emissions have been increasing at about 1% per year.) The model presented here is based on the work by Law et al. (1993). The objective of the modeling effort is to reproduce the two peaks in the methane flux typically noted under field conditions.

$\text{Mass} := 50 \quad n := 0.3 \quad \text{Vel} := 1.1 \quad b := 5 \quad \lambda := 0.002 \quad t := 10$

$\text{Dxx} := 10 \quad \text{Dyy} := 1 \quad j := 0,1 .. 25 \quad k := 0,1 .. 20$

$x_j := 5 + 10 \cdot j \qquad y_k := -100 + 10 \cdot k$

$$C(x,y) := \frac{\text{Mass} \cdot 1000 \cdot e^{\left[\dfrac{-\left[(x - \text{Vel} \cdot t)^2\right]}{4 \cdot \text{Dxx} \cdot t \cdot 30} - \dfrac{y^2}{4 \cdot \text{Dyy} \cdot t \cdot 30}\right]}}{4 \cdot \pi \cdot b \cdot t \cdot 30 \cdot \sqrt{\text{Dxx} \cdot \text{Dyy}}} \cdot e^{(-\lambda \cdot t)}$$

$\text{Conc}_{j,k} := C\left(x_j, y_k\right)$

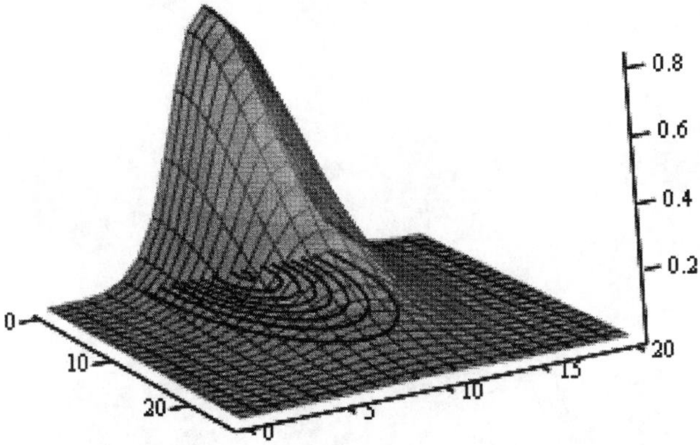

**Figure 9.19** Contaminant transport visualization in Mathcad®.

While the production of methane and its transport to the atmosphere is a complex process, simplifying assumptions can be made to include the most important mechanisms and keep the model reasonably simple. Accordingly, the following assumptions are made in this example: methane is generated from two sources of carbon—carbon initially in the soil and carbon provided by the plants—with the same biokinetic rate according to Monod's kinetic model, methane transport follows simple mass transfer theory, and carbon given off by the plant roots is a function of the physiology of the plant.

Following the above assumptions, the material balance equations can be developed for substrate (S), methanogens (X), and methane (M), respectively:

$$\frac{dS}{dt} = -\left[\frac{kSX}{K_s + S}\right] + K_r A_r f_{(t)}$$

$$\frac{dX}{dt} = -Y\left[\frac{kSX}{K_s + S}\right] - bX$$

$$\frac{dM}{dt} = \left[\frac{kSX}{K_s + S}\right]\left[\frac{1 - Y + 0.8b(K_s + S)}{kS}\right]\frac{16}{59} - K_p M A_r$$

where $f(t)$ is a function modifying the rate of release of substrate by plant roots (known as root exudate) depending on plant physiology. It is set as follows in this example: $f(t) = 0$ if time < 22 days, $f(t) = 0.8$ if time > 22 days and < 45 days, $f(t) = 1.0$ if time > 45 days and < 58 days, and $f(t) = 0$ if time > 58 days. The other parameters in the model are defined in Table 9.2 along with baseline values.

In the original study, Law et al. (1993) used the dynamic simulation package STELLA® (which is the same as ithink®) to model this phenomenon. In this case, the use of the Extend™ package is illustrated as an alternate software, as shown in Figure 9.20.

## 9.7 MODELING EXAMPLE: CHEMICAL EQUILIBRIUM

In this example, modeling of chemical equilibrium systems is illustrated. Traditionally, modeling of such systems has been done either graphically or by using special purpose computer software packages. These models essentially entail solving a set of linear equations. Here, the ease with which such a model can be developed using TK Solver is illustrated for a closed carbonate system.

Table 9.2 Parameters Used in Methane Emission Model

| Symbol | Variable | Value |
|--------|----------|-------|
| $S$ | concentration of substrate (mg/L) | |
| $X$ | concentration of methanogenic biomass (mg/L) | |
| $M$ | Concentration of methane (mg/L) | |
| $t$ | time (days) | |
| $V$ | volume of reactor (m³) | 0.01 |
| $k$ | rate constant (day⁻¹) | 4.0 |
| $K_s$ | Monod half velocity constant (mg/L) | 9.5 |
| $Y$ | yield constant (mg biomass/mg substrate) | 0.04 |
| $b$ | biomass decay rate (day⁻¹) | 0.032 |
| $K_r$ | Mass transfer coefficient for substrate from root (g/m²-day) | 6.0 |
| $A_r$ | surface area of roots (m²) | 0.01 |
| $K_p$ | mass transfer coefficient for methane from root (m-day) | 1.0 |

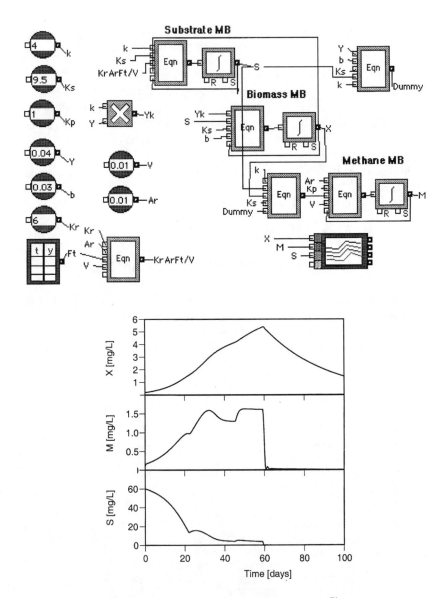

**Figure 9.20** Methane emission modeled in Extend™.

The governing equations are developed as follows from the chemical reactions and the equilibrium constants:

$$H_2CO_3^* \leftrightharpoons HCO_3^- + H^+ \qquad K_1 = \frac{[H^+][CO_3^-]}{[H_2CO_3]} = 10^{-6.3}$$

$$HCO_3^- \leftrightharpoons CO_3^{2-} + H^+ \qquad K_2 = \frac{[H^+][CO_3^{2-}]}{[HCO_3^-]} = 10^{-10.2}$$

$$H_2O \leftrightharpoons OH^- + H^+ \qquad K_w = \frac{[H^+][OH^-]}{[H_2O]} = 10^{-14}$$

The above expressions can be simplified to the following set of simultaneous equations:

$$\log K_1 = \log [H^+] + \log [HCO_3^-] - \log [H_2CO_3]$$

$$\log K_2 = \log [H^+] + \log [HCO_3^{2-}] - \log [HCO_3^-]$$

$$\log K_w = \log [H^+] + \log [OH^-]$$

In addition to the above equations, a total mass balance on carbon can be written as follows:

$$C_T = [H_2CO_3] + [HCO_3^-] + [CO_3^{2-}]$$

The last four equations can be solved to find how a given total carbon mass can dissociate into the three species at various pH values. (Note that $\log [H^+] = -pH$.) The algebraic solution involves polynomial equations requiring a somewhat tedious solution process; however, a trial-and-error process can be readily set up to solve them using a spreadsheet package or an equation solver package.

The model built with TK Solver is presented in Figure 9.21, where the equations are solved at various pH values to plot the speciation diagram. The logarithmic form of the three mass law equations and expressions for pH and the total carbonate concentration are first entered into the *Rule* sheet. The *Variable* sheet is automatically generated by TK Solver, listing all the variables and model parameters under the *Name* column. The model constants $K_1$, $K_2$, and $K_w$, and the total carbonate species, $C_T$, are entered as inputs under the *Input* column. The remaining six lines are defined as *Lists*, with pH as the *Input* list, set by the modeler to vary from 3 to 14 in steps of 1.

Because the solution procedure involves a trial-and-error process, an arbitrary initial guess value has to be provided for one of the unknowns. This is

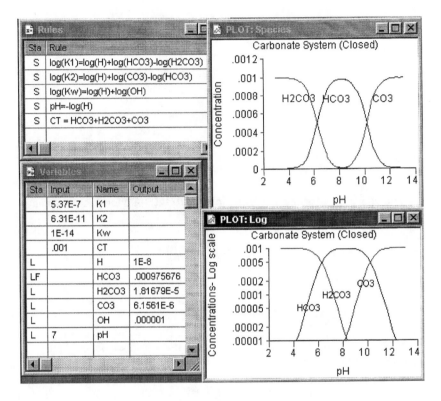

**Figure 9.21** Chemical equilibrium modeled in TK Solver.

done by specifying any one of the unknowns with a guess value. In this example, [HCO₃] was arbitrarily given an initial guess value of 0.001, identified in the *Status* column by the symbol LF, implying that the *List* for that variable had a first value as a guess. When *Run,* the model takes each pH value at a time from the pH *List* and solves all the equations to calculate the concentrations of the other species to generate corresponding *Lists* for them. These lists are then used to produce the plots. When an equation is satisfied during a *Run,* the *Status* column in the *Rules* sheet is filled with the symbol *S* for that equation. For a successful solution, all the equations must be satisfied during a run.

The above model can generate speciation diagrams at any total concentration $C_T$, based on the value entered in the *Variables* sheet under the *Input* column. It can also be readily adapted for any other system by entering appropriate mass laws equations in the *Rules* sheet and corresponding "*K*" values in the *Variables* sheet.

## 9.8 MODELING EXAMPLE: TOXICOLOGICAL EXPOSURE EVALUATION

In this example, a model integrating fate and transport of toxicants with exposure and toxicity assessments is illustrated. Such models can be useful in environmental impact assessment, risk analysis, accidental release management, etc. Consider a situation in which a known amount of a pollutant has been released into an aquifer. Downstream from the release point, a drinking water well draws from this aquifer. It is desired to predict the concentration buildup in animals, for example.

First, the toxicological equations are developed, following the method of Jorgensen (1994), using a three-compartment model consisting of blood, liver, and bones, as an example. A compartment in toxicology is defined as a body component that has uniform kinetics of transformation and transport and whose kinetics are different from those of other components. The modeling framework illustrated can be extended to a greater number of compartments if necessary, and if kinetic data are available for such differentiation.

In the model used here, it is assumed that the intake, $R$, is through the blood compartment (1), part of which is excreted (via the kidneys). The amount transferred to the liver (2) undergoes the metabolic process, modeled by Michaelis-Menten-type kinetics. The transfer to the bones (3) is assimilated. All processes other than metabolism are modeled as first-order processes. These interactions and the kinetics of transfers and transformations are schematically shown in Figure 9.22.

The following equations can be derived from MB across the three compartments:

MB on blood:

$$\frac{dP_1}{dt} = R - k_{3,1}P_1 - k_{2,1}P_1 - k_{0,1}P_1 + k_{1,2}P_2 + k_{1,3}P_3$$

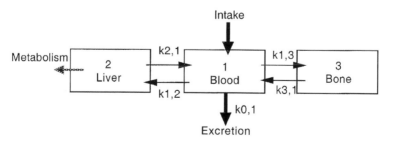

**Figure 9.22** Interactions between compartments.

MB on liver:

$$\frac{dP_2}{dt} = k_{2,1}P_1 - k_{1,2}P_2 - k_r\left(\frac{P_2}{k_m + P_2}\right)$$

MB on bones:

$$\frac{dP_3}{dt} = k_{3,1}P_1 - k_{1,3}P_3$$

The above equations are first used to build the toxicology model using the ithink® dynamic simulation software as shown in Figure 9.23. The kinetic coefficients reported by Jorgensen (1994) are used here and are summarized in Table 9.3, as generated by the ithink® software package. Typical results under a constant intake rate of $R = 1.3$ mg/day are included in Figure 9.23.

The intake of the toxicant by the animal can be established from the toxicant concentration in the aquifer. This concentration, $C$, as a function of time and distance from the source of release can be described by the following generalized, two-dimensional solute transport equation:

$$\frac{\partial C}{\partial t} + v_x\frac{\partial C}{\partial x} + v_y\frac{\partial C}{\partial y} + \lambda C = D_{xx}\frac{\partial^2 C}{\partial x^2} + D_{yy}\frac{\partial^2 C}{\partial y^2}$$

A limited number of analytical solutions have been reported for the above PDE under certain initial and boundary conditions and inputs. Some of these solutions are summarized here:

- pulse input of a reactive, sorptive toxicant, one-dimensional flow:

$$C(x,t) = \frac{M}{\sqrt{4\pi\left(\frac{D_x}{R}\right)t}}\exp\left[-\frac{(x - v_xt)^2}{4\left(\frac{D_x}{R}\right)t}\right]\exp\left(-\frac{kt}{R}\right)$$

- continuous input of nonreactive, sorptive toxicant, one-dimensional flow:

$$C(x,t) = \frac{C_0}{2}\left[\text{erfc}\left(\frac{x - \left(\frac{v_x}{R}\right)t}{2\sqrt{\left(\frac{D}{R}\right)t}}\right) - \exp\left(\frac{\left(\frac{v_x}{R}\right)x}{\left(\frac{D}{R}\right)}\right)\text{erfc}\left(\frac{x + \left(\frac{v_x}{R}\right)t}{2\left(\frac{D}{R}\right)t}\right)\right]$$

- pulse input of nonreactive, sorptive toxicant, two-dimensional flow:

$$C(x,t) = \frac{M'}{4n\pi\sqrt{D_xD_y}}\exp\left[-\left(\frac{(x - v_xt)^2}{4\left(\frac{D_x}{R}\right)t} + \frac{y^2}{4\left(\frac{D_y}{R}\right)t}\right)\right]$$

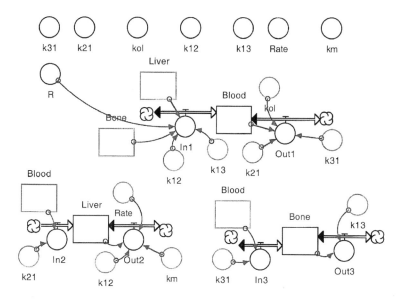

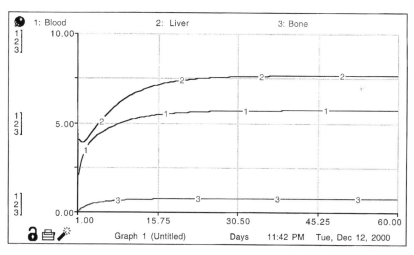

**Figure 9.23** Toxicant in blood, liver, and bone, modeled in ithink®.

The above equations now form the basis of the integrated model. They can be implemented using the ithink® software package with a user-friendly interface as shown in Figure 9.24. The necessary tools to construct such an interface in ithink® are built in. This enables users not familiar with the ithink® environment to run the model under different scenarios.

Table 9.3 Model Equations in ithink®

```
Blood(t) = Blood(t - dt) + (In1 - Out1) * dt
    INIT              Blood = 2
    INFLOWS:          In1 = R + k12*Liver + Bone*k13
    OUTFLOWS:         Out1 = k31*Blood + k21*Blood + kol*Blood

Bone(t) = Bone(t - dt) + (In3 - Out3) * dt
    INIT              Bone = 0
    INFLOWS:          In3 = Blood*k31
    OUTFLOWS:         Out3 - Bone*k13

Liver(t) = Liver(t - dt) + (In2 - Out2) * dt
    INIT              Liver = 4
    INFLOWS:          In2 = Blood•k21
    OUTFLOWS:         Out2 = k12*Liver + Rate*Liver/(km + Liver)

K12 = 0.5
k13 = 0.8
k21 = 0.8
k31 = 0.1
km = 5
kol = 0.4
R = 3
Rate = 1.2
```

## 9.9 MODELING EXAMPLE: VISUALIZATION OF GROUNDWATER FLOW

This example illustrates the use of equation solver-based packages in performing symbolic series calculations involving abstract mathematics to generate visual aids to interpret the results. A groundwater flow in small drainage basins between a water divide and a valley bottom is modeled here as an illustration. The system can be simplified as shown in Figure 9.25.

The governing equation for groundwater flow in this system, derived from Darcy's Law, assuming an isotropic, homogeneous aquifer, is the well-known Laplace equation:

$$\phi_{xx} + \phi_{zz} = 0$$

or

$$\frac{\partial^2 \phi_{(x,z)}}{\partial X^2} + \frac{\partial^2 \phi_{(x,z)}}{\partial z^2} = 0$$

It is desired to develop the flow potential lines and the velocity vectors for the flow field for the system.

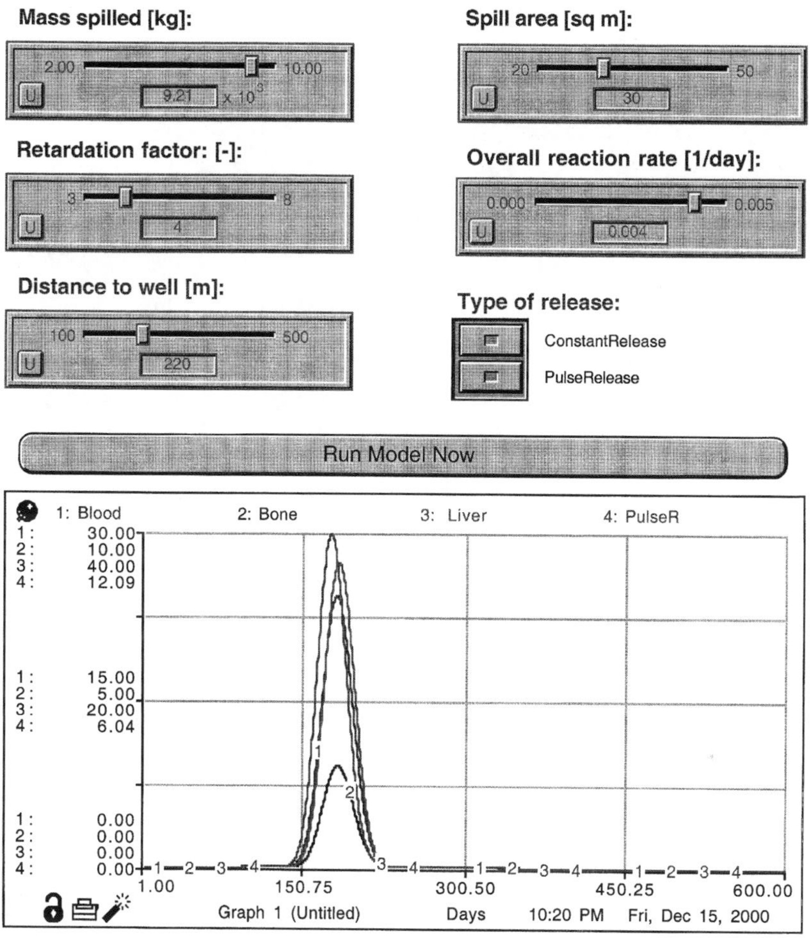

**Mass spilled [kg]:**

2.00 — 10.00
9.21 x 10³

**Spill area [sq m]:**

20 — 50
30

**Retardation factor: [-]:**

3 — 8
4

**Overall reaction rate [1/day]:**

0.000 — 0.005
0.004

**Distance to well [m]:**

100 — 500
220

**Type of release:**

ConstantRelease

PulseRelease

**Run Model Now**

| | 1: Blood | 2: Bone | 3: Liver | 4: PulseR |
|---|---|---|---|---|
| 1: | 30.00 | | | |
| 2: | 10.00 | | | |
| 3: | 40.00 | | | |
| 4: | 12.09 | | | |
| 1: | 15.00 | | | |
| 2: | 5.00 | | | |
| 3: | 20.00 | | | |
| 4: | 6.04 | | | |
| 1: | 0.00 | | | |
| 2: | 0.00 | | | |
| 3: | 0.00 | | | |
| 4: | 0.00 | 1.00    150.75    300.50    450.25    600.00 | | |

Graph 1 (Untitled)        Days        10:20 PM    Fri, Dec 15, 2000

**Figure 9.24** Graphical user interface for toxicant modeled in ithink®.

Four boundary conditions (BCs) are required to solve the above PDE. For the system shown, the following BCs are appropriate:

BC 1: $\phi(x,z_0) = gz_0 + gsx$

BC 2: $\phi_x(0,z) = 0$

BC 3: $\phi_x(L,z) = 0$

BC 4: $\phi_y(x,0) = 0$

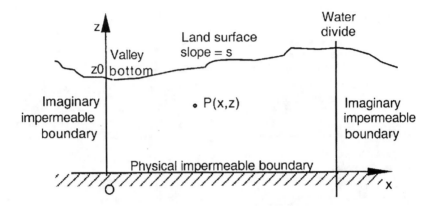

**Figure 9.25** Problem definition for visualization of groundwater flow.

Classical approaches of mathematical calculi can be used to develop the analytical solution for the above PDE. The process begins by assuming a solution of the following form:

$$\phi(x,z) = X(x)Z(z)$$

$$= e^{-kx}(a \cos kx + B \sin kx) + e^{-kz}(M \cos kz + N \sin kz)$$

where A, B, C, and D are arbitrary constants to be found from the BCs. In this case, substituting the above BCs, $B = N = 0$ and $A = M$. On substitution into the PDE, a Fourier series results, which on integration, leads to the final solution for $\phi$:

$$\phi(x,z) = gz_0 + \frac{gsL}{2} - \frac{4gsL}{p^2} \sum_{m=0}^{4} \frac{\cos\left[\dfrac{(2m+1)\pi x}{L}\right] \cosh\left[\dfrac{(2m+1)\pi z}{L}\right]}{(2m+1)^2 \cosh\left[\dfrac{(2m+1)\pi z_0}{L}\right]}$$

The result in the above form is complicated to comprehend or use; however, it can be useful in understanding flow patterns by presenting the relationships between the variables graphically.

In this example, the Mathcad® equation solver-based package is chosen to illustrate how complex equations such as the above can be graphed in several ways for better understanding of the system. Figure 9.26 shows the Mathcad® worksheet, where the model for visualization is encoded.

The same final equation is used to generate the velocity vector plot. Recognizing that the velocity vector is found by differentiating the velocity

$$z0 := 10000 \qquad L := 20000 \qquad s := 0.002 \qquad g := 32 \qquad K := g \cdot z0 + g \cdot s \cdot \frac{L}{2}$$

$$\phi(x,z) := K - \frac{4 \cdot g \cdot s \cdot L}{\pi^2} \sum_{j=0}^{5} \left[ \frac{\cos\left[\frac{(2 \cdot j + 1) \cdot \pi \cdot x}{L}\right] \cdot \cosh\left[\frac{(2 \cdot j + 1) \cdot \pi \cdot z}{L}\right]}{(2 \cdot j + 1)^2 \cdot \cosh\left[\frac{(2 \cdot j + 1) \cdot \pi \cdot z0}{L}\right]} \right]$$

$$N := 10 \qquad p := 0,1 .. N$$

$$q := 0,1 .. N \qquad x_p := 0 + p \cdot \frac{L}{N} \qquad z_q := 0 + q \cdot \frac{z0}{N} \qquad M_{p,q} := \phi\left(x_p, z_q\right)$$

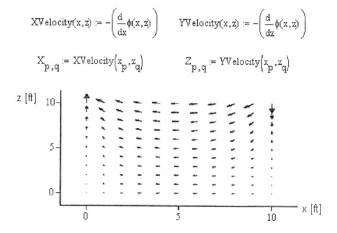

**Figure 9.26** Groundwater flow visualization in Mathcad®.

potential function, the code shown in Figure 9.27 is implemented to determine the components of the velocity vector and to plot the velocity field.

This problem can also be modeled with Mathematica®, as shown in Figure 9.28. First, the model parameters are declared. Then, the equation describing the head is entered using the graphic input palette. The *ContourPlot* routine is then called, where the arguments specify the equation to be plotted and the

$$XVelocity(x,z) := -\left(\frac{d}{dx}\phi(x,z)\right) \qquad YVelocity(x,z) := -\left(\frac{d}{dz}\phi(x,z)\right)$$

$$X_{p,q} := XVelocity\left(x_p, z_q\right) \qquad Z_{p,q} := YVelocity\left(x_p, z_q\right)$$

**Figure 9.27** Groundwater velocity field visualization in Mathcad®.

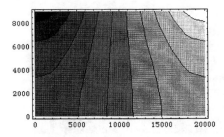

In[1]:= z0 = 10000.0; L = 20000.0; s = 0.0002; g = 32.0;
  K = g*z0 + g*s*L/2;

$$\phi[x, z] = K - \frac{4*g*s*L}{Pi^2} \sum_{m=0}^{5} \frac{Cos\left[\frac{(2m+1)\pi x}{L}\right] Cosh\left[\frac{(2m+1)\pi z}{L}\right]}{(2m+1)^2 Cosh\left[\frac{(2m+1)\pi*z0}{L}\right]} ;$$

ContourPlot[%, {x, 0, 20000}, {z, 0, 9000},
  ContourShading → True, ColorFunction → Automatic,
  AspectRatio -> .6]

**Figure 9.28** Groundwater flow visualization in Mathematica®.

ranges of values of the space coordinates over which the contours are to be generated, and, optionally, the shading and coloring can be made true, with an aspect ratio of the plot. The sign % represents the result found in the previous output line.

## 9.10 MODELING EXAMPLE: AIR POLLUTION—PUFF MODEL

In this example, modeling of an air pollution problem is illustrated. The transport of nonreactive pollutants in the atmospheric system has been well studied. The MB equation for the general case of advective-diffusive transport of gaseous pollutants reduces to the well-known equation:

$$\frac{\partial C}{\partial t} + v_x \frac{\partial C}{\partial x} + v_y \frac{\partial C}{\partial y} + v_z \frac{\partial C}{\partial z} = D_{xx} \frac{\partial^2 C}{\partial x^2} + D_{yy} \frac{\partial^2 C}{\partial y^2} + D_{zz} \frac{\partial^2 C}{\partial z^2}$$

Pasquill (1962) and Gifford (1976) have solved the above PDE by introducing dispersion coefficients, $\sigma_x$, $\sigma_y$, and $\sigma_z$, which are, in turn, related to atmospheric conditions. These formulations are well known and are, therefore, not detailed here. The final result for the time-dependent, two-dimensional, ground-level, spatial concentration profile, for a "puff release" has been reported as follows:

$$C(x,y,t) = \frac{M}{\sqrt{2}\pi^{3/2}\sigma_x\sigma_y\sigma_z} \exp\left\{-\frac{1}{2}\left[\left(\frac{x - Ut}{\sigma_x}\right) + \frac{y^2}{\sigma_2^y}\right]\right\}$$

where $\sigma_x$, $\sigma_y$, and $\sigma_z$ have to be "looked up" from the charts based on wind speed and atmospheric conditions presented by Gifford (1976).

Even though the final result is an algebraic equation, it is not intuitively easy to comprehend. Equation solver-based packages such as MATLAB® and Mathematica® are best suited for analyzing such equations and interpreting the results visually. Because the interactivity is somewhat limited in the Mathematica® environment, the MATLAB® package is used in this example to model and visually analyze this problem by generating various plots.

The first step in implementing the puff model in the computer is to convert the graphical information about the atmospheric conditions to numeric form, either as tables or, preferably, equations. In the case of the puff model, the relationships between horizontal as well as vertical dispersion coefficients, $\sigma_x$, $\sigma_y$, and $\sigma_z$, and downwind distance have been reported as linear in a log-log coordinate system. Thus, correlations for stable, neutral, and unstable conditions can be developed from those plots as follows, where $x$ is in km:

Stable conditions:

$$\sigma_x = \sigma_y = 9x^{0.9}$$

and

$$\sigma_z = 3.6x^{0.82}$$

Neutral conditions:

$$\sigma_x = \sigma_y = 15x^{1.1}$$

and

$$\sigma_z = 19.5x^{0.67}$$

Unstable conditions:

$$\sigma_x = \sigma_y = 90x$$

and

$$\sigma_z = 81.5x^{0.7}$$

With the above equations for the stability curves, the puff model can now be implemented on the computer. The M-File for the puff model is shown in Figure 9.29, where the custom function "Puffs" is coded using the MATLAB® language. The call to this function requires three arguments, which are the three user set variables—mass of chemical released, the wind velocity, and the atmospheric condition, in this example. The code captures the values for these variables from the call and assigns the first argument to

$W$ and the second to $V$, in line 6. The $\sigma_x$, $\sigma_y$, and $\sigma_z$ values are then estimated based on the third argument using a set of "if" statements, in lines 10 to 22. Using these values, the concentration, $C$, at various time values ranging from $T = 1$ to $T = 15$ in steps of three are to be calculated as a function of x and y and plotted. Line 9 sets up this range for $T$ in the subsequent calculations. Line 8 is the code telling MATLAB® to keep the same scale for the axis in the plots to be generated for each $T$ value. Otherwise, MATLAB® automatically sets the scale for different plots. The code is written so that four different types of plots could be generated by activating one of lines 28 to 31.

```
1   function [xx,yy,CC] = Puffs(arg1,arg2,arg3);
2   dx = 1; dy = 1;
3   [x,y] = meshgrid([1:dx:100],[-20:dy:20]);
4   W=arg1; V=arg2;
5   set (gca,'NextPlot','replacechildren')
6   for T = 1:5:20
7       if arg3 == 'U'
8           qy=90*(x/1000).^1;
9           qz=81.5*(x/1000).^0.7;
10          condition='Unstable Conditions';
11      elseif arg3 == 'S'
12          qy=9*(x/1000).^0.9;
13          qz=3.6*(x/1000).^0.82;
14          condition='Stable Conditions';
15      else
16          qy=15*(x/1000).^1.1;
17          qz=19.5*(x/1000).^0.67;
18          condition='Neutral Conditions';
19      end
20      qx = qy;
21      qxqzqy=qz.*qy.*qx;
22      term1=(qxqzqy.^-1)*(W/((2*pi^3)^.5));
23      C=term1.*exp( - ( ((x-V*T).^2)./(qx.^2) + (y.^2).*(qy.^-2) )/2 );
24
25      %surf(x,y,C)
26      %contourf(C,5)
27      %contour(C,5)
28      contour3(C,10)
29      title(condition)
30      hold on
31  end
32  hold off
```

**Figure 9.29** M-file for puff model in MATLAB®.

In the example shown, a three-dimensional contour plot with 10 contours is chosen, and the other three options are commented out by the percent (%) sign. In line 33, the *hold on* command tells MATLAB® to keep the settings the same for all the calculations specified under the *for* statement. To run this code, the user has to type in the call specifying the name of the M-File with the arguments for MATLAB® to execute the code in that file. The resulting plots in Figure 9.30 show the spatial distribution of the pollutant at the ground level at various times, as the puff is carried away by the wind.

This example illustrates some of the mechanics involved in running a MATLAB® code interactively. However, for users not familiar with the MATLAB® environment, this may not be intuitive and easy to adapt. For example, accessing the M-File or modifying it to pick the desired type of plot, or changing the grid sizes, etc., requires some familiarity with the MATLAB® environment. MATLAB® includes tools to create a user-friendly GUI through which a novice can intuitively run MATLAB® files interactively, by clicking buttons and entering inputs through dialog boxes. Such an interface will be included in the next example, where the plume model is illustrated.

## 9.11 MODELING EXAMPLE: AIR POLLUTION—PLUME MODEL

In this example, the computer implementation of the plume model to predict spatial concentrations resulting from a continuous source of air pollutant is illustrated. Again, the well-known Gaussian Dispersion Model and its solution proposed by Pasquill and Gifford are used here. The objective is to develop a model incorporating a user-friendly interface for novices to run the model under various scenarios and generate plots to aid in visual appreciation and analysis of the problem.

The following equations describe the steady state, spatial distribution, and ground-level concentrations of nonreactive gaseous or aerosol (diameter < 20 μm) pollutants emitted by a stack of effective height, $H$:

Spatial distribution:

$$C(x,y,z) = \frac{M}{2\pi\sigma_y\sigma_z U}\left[\exp\left\{-\frac{y^2}{2\sigma_y^2} - \frac{(z-H)^2}{2\sigma_z^2}\right\} + \exp\left\{-\frac{y^2}{2\sigma_y^2} - \frac{(z+H)^2}{2\sigma_z^2}\right\}\right]$$

Ground-level distribution ($z = 0$):

$$C(x,y,z) = \frac{M}{\pi\sigma_y\sigma_z U}\left[\exp\left\{-\frac{y^2}{2\sigma_y^2} - \frac{H^2}{2\sigma_z^2}\right\}\right]$$

where $M$ is the mass rate of emission, $U$ is the mean wind velocity in the $x$ direction and $\sigma_y$ and $\sigma_z$ are the horizontal and vertical dispersion coefficients.

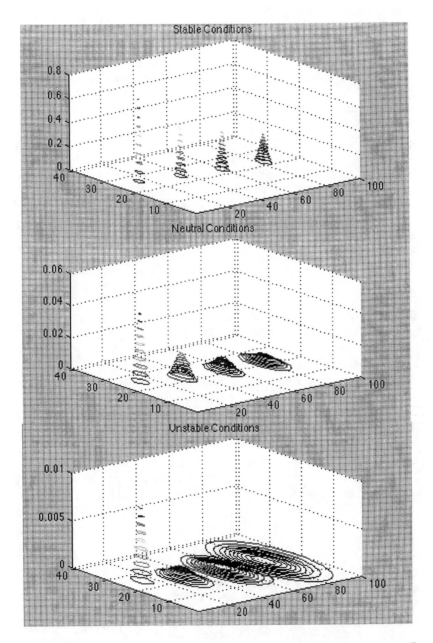

**Figure 9.30** Three-dimensional contours for stable, neutral, and unstable conditions in MATLAB®.

Important assumptions behind these equations are that they are applicable in the above form only to flat terrains, the pollutant is conservative, and the wind velocity is constant.

The effective height of the stack, $H$, can be found from its physical height, $h$, and the plume rise, $\Delta h$, by the following equation:

$$H = h + \Delta h = h + \frac{v_s d}{U}\left[1.5 + 2.68 \times 10^{-3} p_{atm.}\left(\frac{T_s - T_{atm.}}{T_s}\right)d\right]$$

where $v_s$ is the stack gas velocity, $d$ is the stack inside diameter, $p_{atm.}$ and $t_{atm.}$ are the atmospheric pressure and temperature, and $T_s$ is the temperature of the stack gases. The measured wind velocity, $U_0$, at the height of the anemometer, $h_0$, may have to be corrected for stack height, $h$, using the following equation:

$$U = U_0\left(\frac{h}{h_0}\right)^k$$

where the exponent $k$ is often taken as $1/7$.

The dispersion coefficients depend on the downwind distances and the stability of the atmosphere, and they have to be read off plots. The stability conditions have to be established based on wind speed and weather conditions according to the following table:

| Wind Speed (m/s) | Day | | | Night | |
|---|---|---|---|---|---|
| | Strong | Medium | C | Cloudy | Calm and Clear |
| <2 | A | A-B | B | | |
| 2 to 3 | A-B | B | C | E | F |
| 3 to 5 | B | B-C | C | D | E |
| 5 to 6 | C | C-D | D | D | D |
| >6 | C | D | D | D | D |

As part of this example, the curve-fitting feature available in MATLAB® is also illustrated here in transforming the original plots of Pasquill (1962) and Gifford (1976) into equations for integrating into the rest of the model. (Originally, these plots were developed by fitting curves to experimentally measured data, which is a standard procedure in analyzing experimental data and developing mathematical models from physical models as discussed in Chapter 1. Here, that process is mimicked by obtaining the "data" off the charts and re-creating the curves and equations to approximate them.)

The MATLAB® package is used first to develop polynomial equations to characterize the stability curves. The equations obtained from this curve-fitting exercise are summarized here:

Dispersion Coefficients

| Condition | Vertical | | Horizontal | |
|:---:|:---:|:---:|:---:|:---:|
| | $\sigma_z$ | $r^2$ | $\sigma_y$ | $r^2$ |
| A | $0.003x^{1.77}$ | 0.97 | $0.422x^{0.89}$ | 0.99 |
| B | $0.016x^{1.34}$ | 0.98 | $0.292x^{0.89}$ | 0.98 |
| C | $0.201x^{0.81}$ | 0.99 | $0.198x^{0.90}$ | 0.99 |
| D | $0.275x^{0.66}$ | 0.99 | $0.135x^{0.90}$ | 0.99 |
| E | $0.156x^{0.69}$ | 0.99 | $0.107x^{0.89}$ | 0.99 |
| F | $0.145x^{0.68}$ | 0.99 | $0.078x^{0.89}$ | 0.99 |

The above equations are integrated to develop various plots to visualize the results. First, a simple script is written and saved as an M-File by the name of, say, *Air3D*. The script consists of the following lines: specification for a meshgrid in the $x$-$y$-$z$ space, numerical values for the model parameters, and the equation for spatial distribution $C(x,y,z)$. A call to the MATLAB® built-in routine *Slice* is made, identifying the volumetric data $C(x,y,z)$ to be plotted and specifying the locations of the sections through the plume in the $x$, $y$, $z$ space. After saving the M-File, by typing its name, *Air3D,* in the command window, a volumetric plot is generated with sections drawn through the

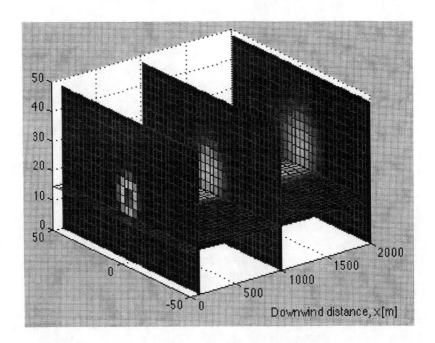

**Figure 9.31** Volumetric plot of plume in MATLAB®.

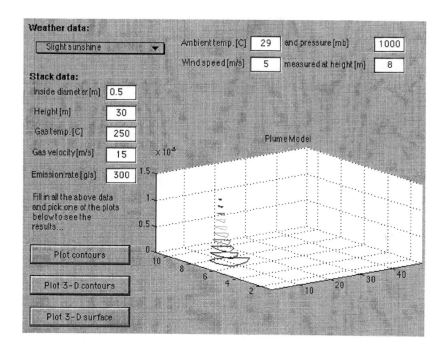

**Figure 9.32** Graphical user interface for plume model in MATLAB®.

plume to illustrate the spread of the plume in the $x$, $y$, and $z$ directions, as shown in Figure 9.31.

A GUI developed using MATLAB®'s built-in tools is shown in Figure 9.32. With this GUI, all the model parameters can be set by the user without having to know anything about the MATLAB® environment. In addition, the GUI also enables the users to visualize the results with three different types of plots. These plots can provide valuable insight into the impacts of air pollution, which cannot be fathomed adequately through the governing equations.

## 9.12 MODELING EXAMPLE: FUGACITY-BASED MODELING

In this example, the fugacity-based modeling approach pioneered by Mackay (1991) and Mackay and Paterson (1981) is adapted to develop a model of the ecosphere. The fugacity concept can be applied to a model "unit world" to gain valuable information about a chemical's behavior in the natural environment. This information can be of significant value in designing a chemical with desired environmental characteristics, managing environmental emissions, ranking chemicals, and environmental policy-making. A basic configuration of the unit world proposed by Mackay consists of the water, soil, air, and sediment compartments. The fugacity concept can be applied at various levels of complexity by including more detailed compartments in the unit world, inter- and intraphase transport and reactive processes, emissions, etc.

Fugacity, $f$, is a measure of a chemical's "escaping tendency" from a phase and is related to its concentration, $C$, in the phase by the equation $f = C/Z$, where $Z$ is the fugacity capacity of the phase. When phases are at equilibrium, their fugacities are equal. For example, in an air-water binary system at equilibrium, $f_w = f_a$, and hence, $C_a/C_w = Z_a/Z_w = K_{a-w}$, the partition coefficient. The $Z$ values for air (1), water (2), soil (3), and sediment (4) compartments can be calculated from the physical-chemical properties of the chemical as follows:

$$Z_1 = \frac{1}{RT}; \quad Z_2 = \frac{1}{H}; \quad Z_3 = K_{s-w}Z_2 = (0.41 Y_{soil} K_{o-w} \rho_{soil})$$

and

$$Z_4 = K_{s-w}Z_2 = (0.41 Y_{sed} K_{o-w} \rho_{sed})$$

where $R$ is the Ideal Gas constant, $T$ is the absolute temperature, $H$ is the chemical's Henry's Constant, $Y$ is the organic content of soil or sediments, $K_{o-w}$ is the chemical's octanol-water partition coefficient, and, $\rho$ is the density of soil or sediment. The fugacity of the system, $f$, at equilibrium is calculated from the following:

$$f = \frac{I}{\Sigma(V_i Z_i k_i + G_i Z_i)} = \frac{E + \Sigma(G_i C_{B,i})}{\Sigma(V_i Z_i k_i + G_i Z_i)}$$

where $V_i$ is the volume of compartment $i$, $G_i$ is the advective flow through compartment $i$, $C_{B,i}$ is the inflow concentration, $k_i$ is the first-order reaction rate constant in compartment $i$, and $E$ is the total emission into the system. The above equations enable concentrations in the four compartments, $C_i$, to be calculated based on the chemical's physical-chemical properties, $H$, $K_{o-w}$, $k_i$; the compartmental characteristics, $V_i$, $Y$, $\rho$, $G_i$, and $C_{B,i}$; and the system temperature, $T$.

In this example, this Level II fugacity model including air, water, soil, and sediment compartments and emissions into the air and water compartments is illustrated. The equations forming the model are all algebraic and can be computerized readily with a spreadsheet package such as Excel®, as shown in Figure 9.33. As can be seen from this figure, in Excel®, the equations have to be set up in a certain order and can be solved only in that order. If the model has to be backsolved, the utility value of the Excel® model will be very limited. Excel®'s *Goal Seek* or *Solver* functions may be used for backsolving, but

Level II Fugacity Calculations:

| | | |
|---|---|---|
| T | K | 300 |
| H | Pa-cu m/mol | 10 |
| Ko-w | mol/hr | 1.00E+04 |
| E | mol/hr | 100 |
| Ysoil | - | 0.02 |
| Ysediment | - | 0.04 |
| ρsoil | kg/cu m | 1500 |
| ρsediment | kg/cu m | 1500 |

| | | Air | Water | Soil | Sediment |
|---|---|---|---|---|---|
| V | cu m | 6.0E+09 | 7.0E+06 | 4.5E+04 | 2.1E+03 |
| Gi | cu m/hr | 1.0E+07 | 1.0E+03 | 0.0E+00 | 0.0E+00 |
| CBi | mol/cu m | 1.0E-06 | 1.0E-02 | | |
| Gi x CBi | mol/hr | 1.0E+01 | 1.0E+01 | 0.0E+00 | 0.0E+00 |
| Σ Gi x Cbi | | 2.0E+01 | | | |
| E + ΣGi x Cbi | mol/hr | 1.2E+02 | | | |
| Zi | mol/cu m-Pa | 0.0004 | 0.10 | 12.30 | 24.60 |
| Vi x Zi | mol/Pa | 2.4E+06 | 7.0E+05 | 5.5E+05 | 5.2E+04 |
| ki | 1/hr | 0 | 0.001 | 0.01 | 0.0001 |
| Vi x Zi x ki | mol/Pa-hr | 0 | 700 | 5535 | 5.166 |
| Gi x Zi | mol/Pa-hr | 4009 | 100 | 0.0 | 0.0 |
| Σ | mol/Pa-hr | 4009 | 800 | 5535 | 5.2 |
| Σ(VixZixki+GixZi)mol/Pa-hr | | 1.0E+04 | | | |
| f | Pa | 1.2E-02 | | | |
| C = f x Zi | mol/cu m | 4.6E-06 | 1.2E-03 | 1.4E-01 | 2.9E-01 |
| Amount | mol | 2.8E+04 | 8.1E+03 | 6.4E+03 | 6.0E+02 |
| Percent | % | 65% | 19% | 15% | 1% |

**Figure 9.33** Level II fugacity model in Excel®.

these two features can backsolve for only one variable that is specified as the target cell.

This simple spreadsheet can be used for preliminary evaluation of the fate of chemicals in the environment under several simplifying assumptions. It can also generate a series of graphs as shown in Figure 9.34 to illustrate the chemical's environmental behavior.

An ability to backsolve for more than one variable at a time can be of significant value in this problem. Some of the situations that can be studied with models capable of backsolving include finding the maximum emission that

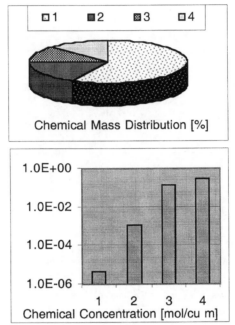

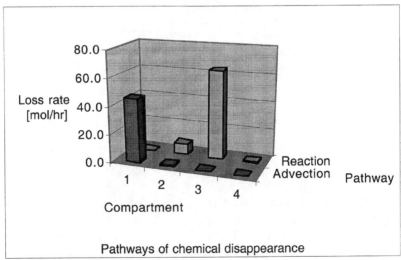

**Figure 9.34** Results of Level II fugacity model in Excel® (key for compartments: (1) Air; (2) Water; (3) Soil; (4) Sediment).

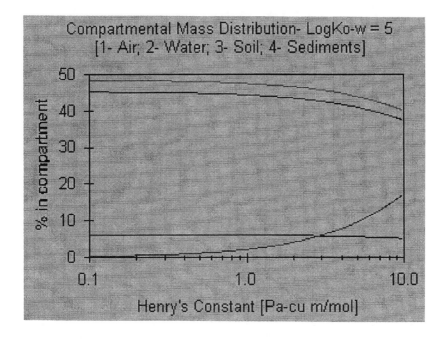

**Figure 9.35** Results of Level II fugacity model in TK Solver.

can be tolerated without exceeding some specified concentrations in the compartments, estimating the range of physical and chemical properties of new chemicals that can meet desired environmental standards in the various compartments, determining or assessing the impacts of waste minimization efforts, and so on. It is therefore fitting to use the TK Solver package to model this problem, whereby its backsolving ability can be utilized to the maximum. A plot generated by the TK Solver model illustrating the compartmental distribution of a chemical at various Henry's Constant values and a fixed log $K_{ow}$ value of five is shown in Figure 9.35.

## 9.13 MODELING EXAMPLE: WELL PLACEMENT AND WATER QUALITY MANAGEMENT

This example illustrates the use of the potential theory in well placement and water quality control. A production well is to be located in an unconfined aquifer adjacent to a river. The current groundwater flow is perpendicular to the river flow. The total dissolved solids (TDS) measurement in the river is 1200 mg/L, while that in the aquifer recharging the river is 300 mg/L. The well has to be placed as close as possible to the river, but it should meet a water quality standard of 500 mg/L TDS. The proposed pumping rate is 10 gpm (20,000 ft³/day), while the average aquifer flux is $U = 5$ ft²/day.

Stream function and the velocity potential functions developed from the potential theory can be used in modeling this scenario. A coordinate system, with origin at $O$, is first chosen, as shown in Figure 9.36.

To create a constant potential boundary along the river, an "image" injection well is used on the opposite side of the river. Using superposition, the potential at any point $P$ can be expressed as follows:

$$\phi = \phi_0 + Ux + \frac{Q}{2\pi} \ln (r_1) - \frac{Q}{2\pi} \ln (r_2)$$

The flow field velocity in the $x$-direction, $Ux$, along the river (where $x = 0$), for various locations of the well can now be found from the following:

$$(U_x)_{(0,y)} = -\left(\frac{\partial \phi}{\partial x}\right)_{(0,y)}$$

$$= -U - \frac{Q}{2\pi}\left(\frac{-L}{L^2 + y^2}\right) + \frac{Q}{2\pi}\left(\frac{L}{L^2 + y^2}\right)$$

When the above result equals zero, or in other words, $U_x$ along the river is zero, all the flow into the well will come from the aquifer. Setting the above to zero yields the following condition:

$$\left(\frac{y}{L}\right)^2 = \frac{Q}{\pi L U} - 1 = \lambda - 1$$

which can have one of the following results:

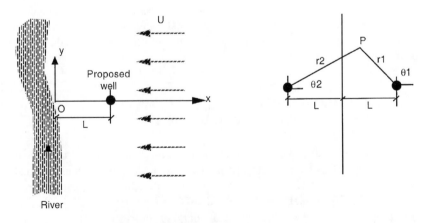

**Figure 9.36** Problem definition for well location.

two roots, at $\left(\dfrac{y}{L}\right) = t\sqrt{\pm\lambda - 1}$ $\qquad$ if $\lambda > 1$

one root, at $y = 0$ $\qquad$ if $\lambda = 1$

no real roots $\qquad$ if $\lambda < 1$

The last case represents no flow from the river toward the proposed well, with the well receiving 100% of the water from the aquifer. Using the given data, $\lambda = 1$ when $L = Q/\pi U = 1272$ ft. Therefore, if the well is located, say, 1300 ft from the river, the TDS in the well output will be 300 mg/L. However, this is well below the water quality standard of 500 mg/L. The well can be moved closer to the river and receive a fraction of its capacity from the river. This fraction of flow rate, $f_r$, can be found from a simple mass balance:

$$500 \text{ mg/L} = f_r \times 1200 \text{ mg/L} + (1 - f_r) \times 300 \text{ mg/L}$$

giving $f_r = 2/9$. The value of $L$ to achieve this split of flow is shown in Figure 9.37 and can be found as follows.

The stream function along the river is found by setting $x = 0$, and $\theta 1 = \pi - \theta 2$:

$$\psi(x = 0,y) = Uy + \frac{Q}{2} - \frac{Q}{\pi}\tan^{-1}\left(\frac{y}{L}\right)$$

and the stream function at $y = L\sqrt{(\lambda - 1)}$ is as follows:

$$\psi(x = 0, L\sqrt{\lambda - 1}) = UL\sqrt{\lambda - 1} + \frac{Q}{2} - \frac{Q}{\pi}\tan^{-1}(\sqrt{\lambda - 1})$$

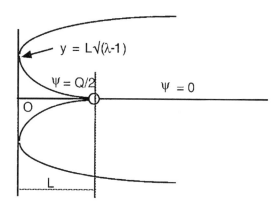

**Figure 9.37** Definition diagram for well location.

Hence,

$$f_r = \frac{2}{9} = \frac{\frac{Q}{\pi}\tan^{-1}\sqrt{\lambda-1} - UL\sqrt{\lambda-1}}{Q/2} = \frac{2}{\pi}\left(\tan^{-1}\sqrt{\lambda-1} - \frac{\sqrt{\lambda-1}}{\lambda}\right)$$

The above can be solved by trial and error to find $\lambda$ as = 2.3, and hence, $L$ from $L = Q/(\pi\, U\lambda)$ as $L = 553$ ft. Hence, the well can be as close as 600 ft to the river and meet the water quality standard.

A plot of the stream function can illustrate the above concepts visually. The composite stream function for this problem can be formulated by adding the stream functions for a uniform flow field representing the aquifer, a source representing the proposed well, and a sink representing the image well.

$$\psi = Uy + \frac{Q}{2\pi}\theta_1 - \frac{Q}{2\pi}\theta_2$$

$$= Uy + \left(\frac{Q}{2\pi}\right)\tan^{-1}\left(\frac{y}{x-L}\right) - \left(\frac{Q}{2\pi}\right)\tan^{-1}\left(\frac{y}{x+L}\right)$$

The Mathematica® model of this problem to generate the streamlines and the velocity potentials is shown in Figure 9.38 for $L = 600$ ft.

As a comparison, the streamlines, velocity potentials, and combined plots for the "safe" case with $\lambda = 1$ or $L = 1300$ ft are shown in Figure 9.39.

```
In[78]:= u = 1; r = 200; q = 500; l = 600;
        ContourPlot[
          ψ[x, y] = u*y + q*ArcTan[y/(x-1)] - q*ArcTan[y/(x+1)],
          {x, 0, 2000}, {y, -1000, 1000}, PlotPoints → 50,
          Contours → 30, ContourLines → True,
          ContourShading → False]

In[79]:= u = 1; r = 200; q = 500; l = 600;
        ContourPlot[
          φ[x, y] = u*x + q*Log[((x-1)^2 + y^2)] -
            q*Log[((x+1)^2 + y^2)], {x, 0, 2000},
          {y, -1000, 1000}, PlotPoints → 50, Contours → 30,
          ContourLines → True, ContourShading → False]
```

**Figure 9.38a** Script for stream lines and velocity potentials in Mathematica®.

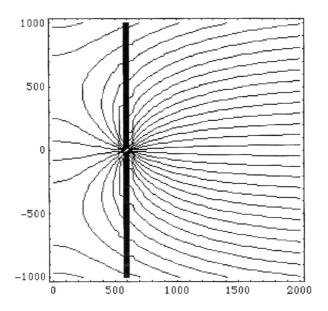

**Figure 9.38b** Stream lines generated by Mathematica®; $L = 600$ ft.

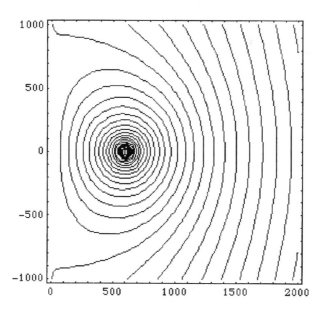

**Figure 9.38c** Velocity potential lines generated by Mathematica®; $L = 600$ ft.

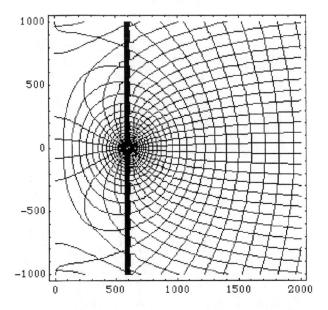

**Figure 9.38d** Stream lines and velocity potential lines in Mathematica®; $L = 600$ ft.

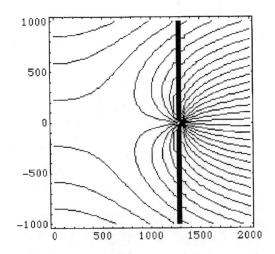

**Figure 9.39a** Stream lines generated by Mathematica®; $L = 1300$ ft.

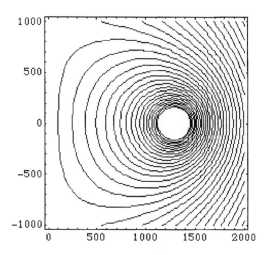

**Figure 9.39b** Velocity potential lines generated by Mathematica®; $L = 1300$ ft.

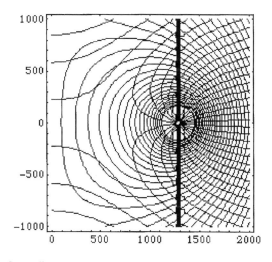

**Figure 9.39c** Stream lines and velocity potential lines generated by Mathematica®; $L = 1300$ ft.

# Bibliography

Bedient P. B., Rifai, H. S., and Newell, C. J. (1994) *Groundwater Contamination*, Prentice Hall, New Jersey.

Bedient P. B., Rifai, H. S., and Newell, C. J. (1999) *Groundwater Contamination*, 2nd ed., Prentice Hall, New Jersey.

Cellier, F. E. (1991) *Continuous System Modeling*, Springer-Verlag, New York.

Chapra, S. C. (1997) *Surface Water-Quality Modeling*, McGraw-Hill, New York.

Charbeneau, R. J. (2000) *Groundwater Hydraulics and Pollutant Transport*, Prentice Hall, New Jersey.

Clark, M. M. (1996) *Transport Modeling for Environmental Engineers and Scientists*, John Wiley & Sons, New York.

Cutlip, M. B., Hwalek, J. J., Nutall, E., Shacham, M., Brule, J., Widmann, J., Han, T., Finlayson, B., Rosen, E. M., and Taylor, R. A collection of 10 numerical problems in chemical engineering solved by various mathematical software packages, *Comput. Appl. Eng. Educ.*, 7, 169–180, 1998.

Davies, G. A. (1985) *Mathematical Methods in Engineering*, John Wiley & Sons, Suffolk, UK.

Dick, S., Riddle, A., and Stein, D. (1997) *Mathematica in the Laboratory*, Cambridge University Press, Cambridge, U.K.

El Shayal, I. Solving differential equations on a spreadsheet—Part I, *Chem. Eng.*, 7, 149–150, 1990a.

El Shayal, I. Solving differential equations on a spreadsheet—Part II, *Chem. Eng.*, 7, 153–154, 1990b.

Fetter, C. W. (1999) *Contaminant Hydrogeology*, Prentice Hall, New Jersey.

Ford, A. (1999) *Modeling the Environment*, Island Press, Washington, D.C.

Gifford, F. A. Turbulent diffusion typing schemes: A review, *Nucl. Safety*, 17, 68–86, 1976.

Goeddertz, J. G., Matsumoto, M. R., and Webber, A. S. Offline bioregeneration of granular activated carbon, *ASCE J. Env. Eng.*, 114, 1063–1075, 1988.

Gottfried, B. S. (2000) *Spreadsheet Tools for Engineers*, McGraw-Hill Co., New York.

Hardisty, J., Taylor, D. M., and Metcalfe, S. E. (1993) *Computerised Environmental Modeling—A Practical Introduction Using Excel*, John Wiley & Sons, U.K.

Hermance, J. F. (1998) *A Mathematical Primer on Groundwater Flow*, Prentice Hall, New Jersey.

James, A. (1993) *An Introduction to Water Quality Modeling*, John Wiley & Sons, New York.

Johnson, P. C., Stanley, C. C., Kemblowski, M. W., Byers, D. L., and Colthart, J. D., A practical approach to the design, operation, and monitoring of *in situ* soil-venting systems, *Ground Water Monitoring Rev.*, 10, 159–178, 1990.

Jorgensen, S. E. (1994) *Fundamentals of Ecological Modeling*, Elsevier, Amsterdam.

Kharab, A. Spreadsheets solve boundary-value problems, *Chem. Eng.*, 7, 145–148, 1988.

Laughlin, D. K., Edwards, F. G., Egemen, E., and Nirmalakhandan, N. Sequencing batch reactor treatment of high COD effluent from a bottling plant, *ASCE J. Env. Eng.*, 125, 285–289, 1999.

Law, V. J., Johnson, N. L., Oyefodun, A., and Bhattachaya, S. K. Modeling methane emissions from rice soils, *Environmental Software*, 8, 197–207, 1993.

Levenspiel, O. (1972) *Chemical Reaction Engineering*, 2nd ed., John Wiley & Sons, New York.

Lincoff A. H. and Gossett, J. M. (1984) The determination of Henry's Constants for volatile organics by equilibrium partitioning in closed systems. In: *Gas Transfer at Water Surfaces*, Eds., Brutsaert, W. and Jirka, G. H. D. Reidel Publishing, Boston, pp. 17–25.

Logan, B. E. (1999) *Environmental Transport Processes*, John Wiley & Sons, New York.

Lyman, W. J., Reehl, W. F., and Rosenblatt, D. H. (1982) *Handbook of Chemical Property Estimation Methods*, McGraw-Hill, New York.

Mackay, D. (1991) *Multimedia Environmental Models*, Lewis Publishers, Michigan.

Mackay, D. and Paterson, S. Calculating fugacity, *Environ. Sci. & Technol.*, 15, 1006–1014, 1981.

Mackay, D. and Paterson, S. Fugacity revisited, *Environ. Sci. & Technol.*, 16, 654a–660a, 1982.

Mackay, D. and Paterson, S. Evaluating the multimedia fate of organic chemicals: A Level III fugacity model, *Environ. Sci. & Technol.*, 25, 427–436, 1991.

Mackay, D., Paterson, S., and Schroeder, W. H. Model describing the rates of transfer processes of organic chemicals between atmosphere and water, *Environ. Sci. & Technol.*, 20, 810–816, 1981.

Marchand, P. (1999), *Graphics and GUIs with Matlab*, CRC Press, Boca Raton, Florida.

Murthy, D. N. P., Page, N. W., and Rodin, E. Y. (1990) *Mathematical Modeling*, Pergamon Press, Exeter, U.K.

Nemerow, N. L. (1991) *Stream, Lake, Estuary, and Ocean Pollution*, van Nostrand Reinhold, New York.

Nirmalakhandan, N., Lee, Y. H., and Speece R. E. Optimizing oxygen absorption and nitrogen desorption in packed towers, *Aquacultural Eng.*, 7, 221–234, 1988.

Nirmalakhandan, N., Jang, W., and Speece, R. E. Counter-current cascade air-stripping for removal of low volatile organic contaminants, *Water Res.*, 24, 615, 1990.

Nirmalakhandan, N., Speece, R. E., and Peace G. L. Cascade air-stripping—A new technology to meet new regulations, *Chem. Eng. Comm.*, 102, 47, 1991a.

Nirmalakhandan, N., Jang, W., and Speece, R. E. Cascade air-stripping—Pilot and prototype scale experience, *J. Env. Eng., ASCE*, 117, 788, 1991b.

Nirmalakhandan, N., Jang, W., and Speece, R. E. Removal of 1,2 dibromo-3-chloro-propane by cascade air-stripping, *J. Env. Eng., ASCE,* 118, 226, 1992a.

Nirmalakhandan, N., Jang, W., and Speece, R. E. Cascade air-stripping: Techno-economic evaluation of a novel groundwater treatment process, *Ground Water Monitoring Rev.,* Spring, 100, 1992b.

Nirmalakhandan, N., Speece, R. E., Peace, G. L., and Jang, W. Operation of counter-current air-stripping towers at higher loading rates, *Water Res.,* 27, 807, 1993.

Palm III, W. J. (2001) *Introduction to Matlab 6 for Engineers,* McGraw-Hill, New York.

Pasquill, F. (1962) *Atmospheric Diffusion,* Van Nostrand, London, U.K.

Pritchard, P. J. (1999) *Mathcad—A Tool for Engineering Problem Solving,* McGraw-Hill, New York.

Schnoor, J. L. (1996) *Environmental Modeling,* John Wiley & Sons, New York.

Schnoor, J. L. and O'Connor, D. J. A steady state eutrophication model for lakes, *Water Res.,* 14, 1651–1655, 1980.

Shacham, M and Cutlip, M. B. A comparison of six numerical software packages for educational use in chemical engineering curriculum, *Computers in Edu. J.,* IX, 9, 1999.

Shoup, T. E. (1979) *A Practical Guide to Computer Methods for Engineers,* Prentice-Hall, New Jersey.

Speece, R. E., Nirmalakhandan, N. N., and Lee, Y. H. Design for high purity oxygen absorption and nitrogen stripping for fish culture, *Aquacultural Eng.,* 7, 201–210, 1988.

Thibodeaux, L. J. (1996) *Environmental Chemodynamics,* John Wiley & Sons, New York.

Thomann, R. V. and Mueller, J. A. (1987) *Principles of Surface Water Quality Modeling and Control,* Harper Collins Publishers, New York.

Treybal, R. E. (1980) *Mass Transfer Operations,* 3rd ed., McGraw-Hill, New York.

Watts, R. J. (1997) *Hazardous Wastes: Sources, Pathways, Receptors,* John Wiley & Sons, New York.

Webber, W. J. Jr. (1972) *Physicochemical Processes for Water Quality Control,* Wiley Interscience, New York.

Webber, W. J. and DiGiano, F. A. (1996) *Process Dynamics in Environmental Systems,* John Wiley & Sons, New York.

# Index

Printed in the United States
by Baker & Taylor Publisher Services